THE WORTH OF WATER

THE WORTH OF WATER

Technical Briefs on Health, Water and Sanitation

With an introduction
by John Pickford

NOTE
The Technical Briefs in this book have all been edited and most of them written by the staff of the Water, Engineering and Development Centre (WEDC). WEDC is a self-funding unit within Loughborough University of Technology devoted to training, research, consultancy and other activities related to the planning, provision, operation and maintenance of water supplies, sanitation and physical infrastructure in developing countries. WEDC is especially concerned with all that is appropriate for rural areas and low-income communities.

Published by Practical Action Publishing,
The Schumacher Centre for Technology and
Development, Bourton on Dunsmore, Rugby, Warwickshire CV23 9QZ, UK.
www.practicalactionpublishing.org

Reprinted 1998, 2007, 2010

ISBN 978 185339 069 2

Printed and bound in Great Britain by
CPI Antony Rowe, Chippenham and Eastbourne

CONTENTS

INTRODUCTION

WE ARE TOLD THAT many subscribers to the quarterly periodical for fieldworkers, *Waterlines*, look at the centre pages as soon as they receive a new edition. There they find 'Technical Briefs', four pages of concentrated information and advice for field workers and decision-makers. The Technical Briefs have proved so popular and are reported to be so useful, that IT Publications has decided to put together all the Briefs in this book, in a form that can be convenient to fieldworkers, and for reference.

The first thing that may strike the reader is the variety of the topics covered, from soap-making to the engineering design of pipe networks, and from making simple ferro-cement jars to the design of slow sand filtration systems. All deal with practical matters, and because they are brief the authors have had to distill the material and eliminate what is not essential. All are well-illustrated, but here again there is variety. Some authors rely largely on drawings, tables and charts to convey their message; in other Briefs the authors describe the technology in more detail. Most of the Briefs were written and prepared by the staff of the Water, Engineering and Development Centre (WEDC) at Loughborough University of Technology.

Looking forward beyond the International Drinking Water Supply and Sanitation Decade of 1981-90 to 'Health for all by the Year 2000', massive efforts will be needed to reduce the number of people without adequate water supply and sanitation, or at the very least to avoid too great an increase in the numbers of those who lack these essential services. Even if there is a significant growth in achievement in the 1990s, these efforts will have to continue beyond the year 2000 to keep pace with the increase of population, particularly in urban areas.

Intermediate Technology, the Water, Engineering and Development Centre and many other organizations have been advocating appropriate technology for a quarter of a century. At last these ideas are widely accepted, and a major theme for the 1990s is 'some for all rather than all for some'. Too often in the past, water resources, financial resources, material resources and human resources in the water and sanitation sector have been devoted to works that largely benefit the better off. Too often the aim has been to give some people the piped systems that provide ample quantities of water treated to high standards, and to remove wastes by expensive sewerage networks, taking them for treatment in complex plants.

Now it is realized that there must be a substantial reduction in the cost of services through increased efficiency and through the use of low-cost appropriate technologies. One of the most pressing needs in the sector is providing coverage for growing urban populations. In almost every developing country urban population growth exceeds the national average. Horrific forecasts of the number of mega-cities that will develop in poor countries have been widely publicized. Many of the additional numbers will be accommodated in shanty towns but these fringe or peri-urban districts around growing towns are usually underserviced. So their needs must have priority.

Low-cost (appropriate) technologies have often been thought to be most suitable for rural water supplies, but they also need to be employed for their urban fringes. As far as sanitation is concerned, low-cost technologies are more often appropriate for urban areas than is generally recognized. To be realistic, sewerage is rarely achievable for poor people. The full cost of piped sewerage is beyond the resources available in many middle-income countries as well as those classified as low-income.

Selection of the most appropriate technologies is important, so it is essential that planners, bureaucrats, politicians and other decision-makers have plenty of information about alternatives. Those who are active in the field (whether engineers, health staff or community and development workers) need more technology. They need to be able to advise those who make decisions. They also have to advise or guide and direct those who are actually involved in implementation. Increasingly the local people — individual householders and local community groups — are the main actors and actresses in the implementation of projects and programmes, especially for sanitation. The fieldworkers therefore require details of technology that can be passed on. Many of these workers and many of those who carry out the work have little technical training or experience, so appropriate technology has to be simple. To be effective, publications intended to disseminate technical information to these groups also need to be presented in a non-technical form. The 'Technical Briefs' reproduced in this book, attempt to do precisely this.

I have often said that technology is not enough. For programmes and projects to result in the provision of water supply and sanitation, good (appropriate) technology must go hand in hand with good and appropriate management and other relevant 'software': health education, community participation and human resources development are all vitally important in achieving successful provision of water supplies and sanitation. Moreover, actually providing technology by construction and installation is not enough. Sustainability should have high priority, and this means that the ongoing operation and maintenance of systems must receive at least as much attention as the initial provision of facilities.

John Pickford,
Water, Engineering and Development Centre (WEDC)
Loughborough

1. Household water storage

How much is needed

A family of six in the tropics needs about 30 litres a day of hygienically-safe water for drinking, tooth-cleaning, food preparation and cleaning cooking and eating utensils. Safe water must either be obtained from safe sources or treated.

A 30-day dry period requires about 1,000 litres of stored water. Clean water that is not necessarily bacteriologically safe can be used for other purposes such as bathing, washing clothes and latrine cleaning.

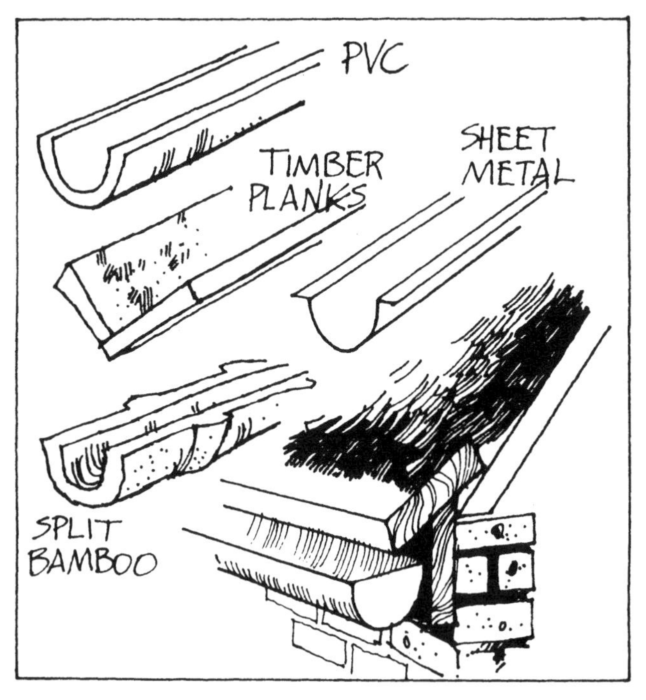

Rain is the safest source

Rainwater can be collected from roofs in gutters made from wood or metal. Divert the first few minutes of flow to a drain, as it will carry the dust and debris that have accumulated on the roof since the last rain. Allow the main flow to pass through a cloth or wire screen.

Store water in clean vessels fitted with covers.

Take water from the storage vessels with a tap if possible. If a dipper is used, keep the dipper clean. Store it inside the vessel if possible.

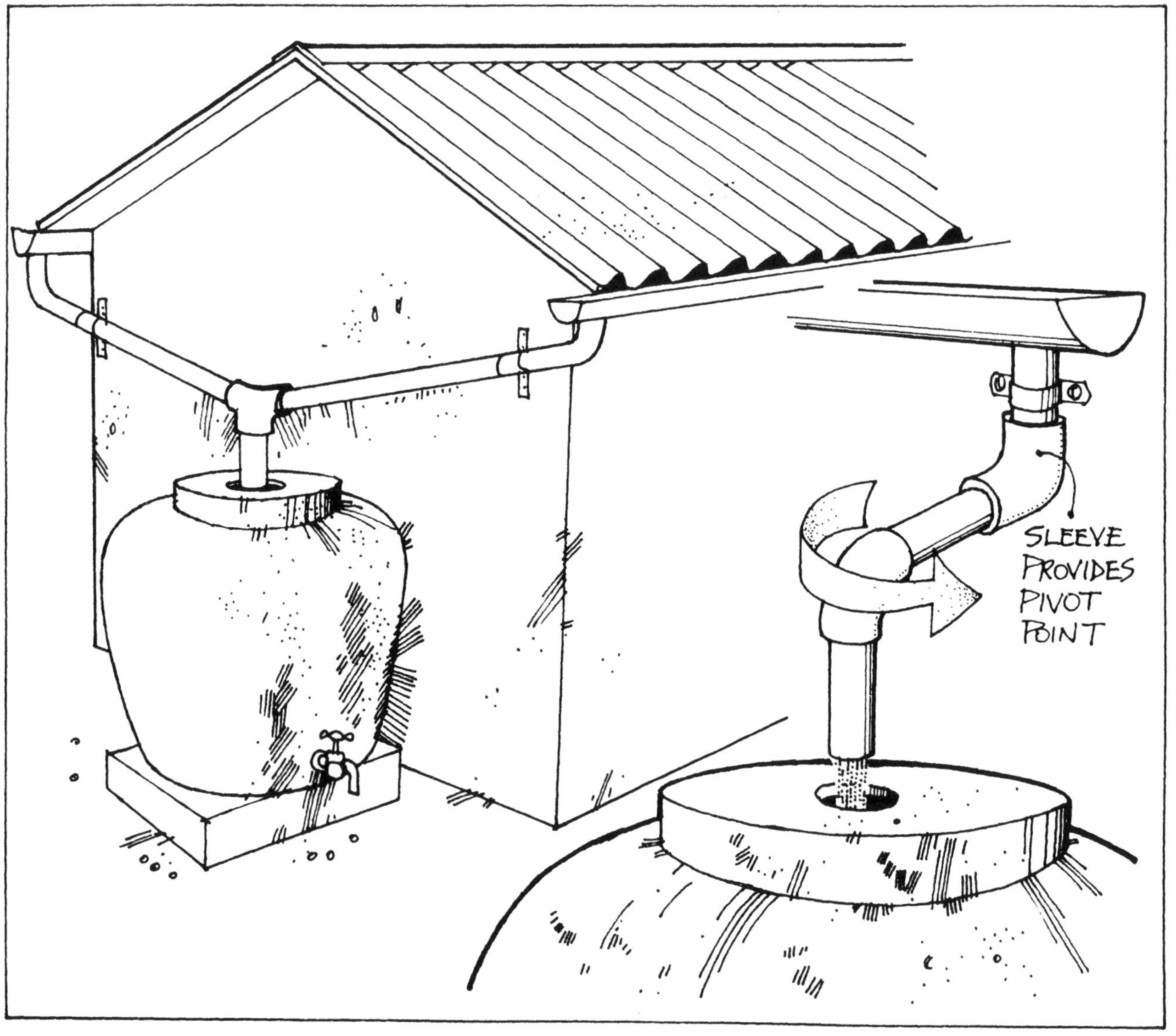

Building un-reinforced mortar jars

1. For a 250 litre jar obtain 1 bag (50kg) sharp sand, ½ a bag (25kg) cement, and water.

When mixing the plaster use as little water as possible, but enough to allow the mixture to hold together. If a lump of the mixture retains its shape when thrown to the ground and does not spread with the impact, it is about the right consistency.

2. Cast the base plate. First lay wrapping paper on smooth ground. Place a 600mm diameter ring made of metal strip 20mm wide on the paper and make up enough mortar to fill it. Allow the base to dry for one day under wet hessian and then remove the ring, which can be used again.

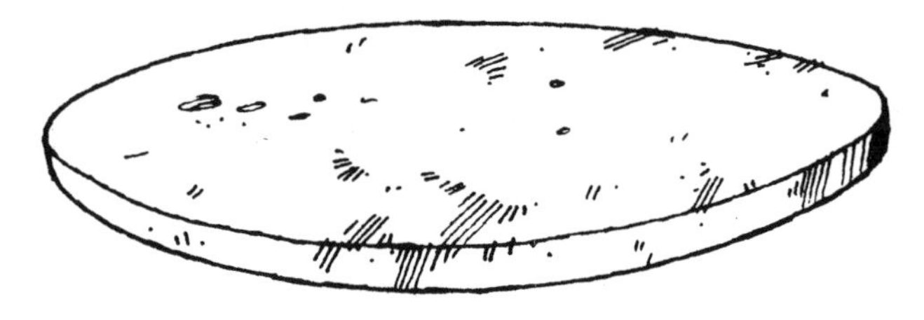

3. The mould for the jar can be made by sewing two pieces of hessian together to form a sleeve.

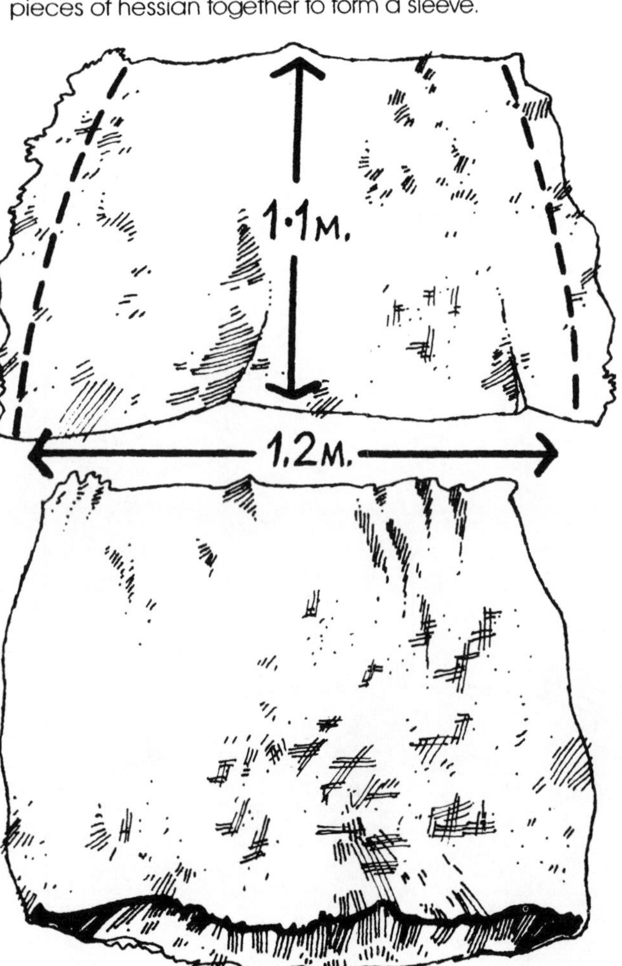

4. Set up the sleeve on the prepared base, tucking the hessian well in at the bottom, and fill it with sand, sawdust, or paddy husk. Fold in the top and tie.

5. Place a second metal ring of 400mm diameter, made of 70-80mm wide strip on the top of the bag. This will form the neck. Spray the bag with plenty of water and make it into a smooth rounded shape with a piece of wood.

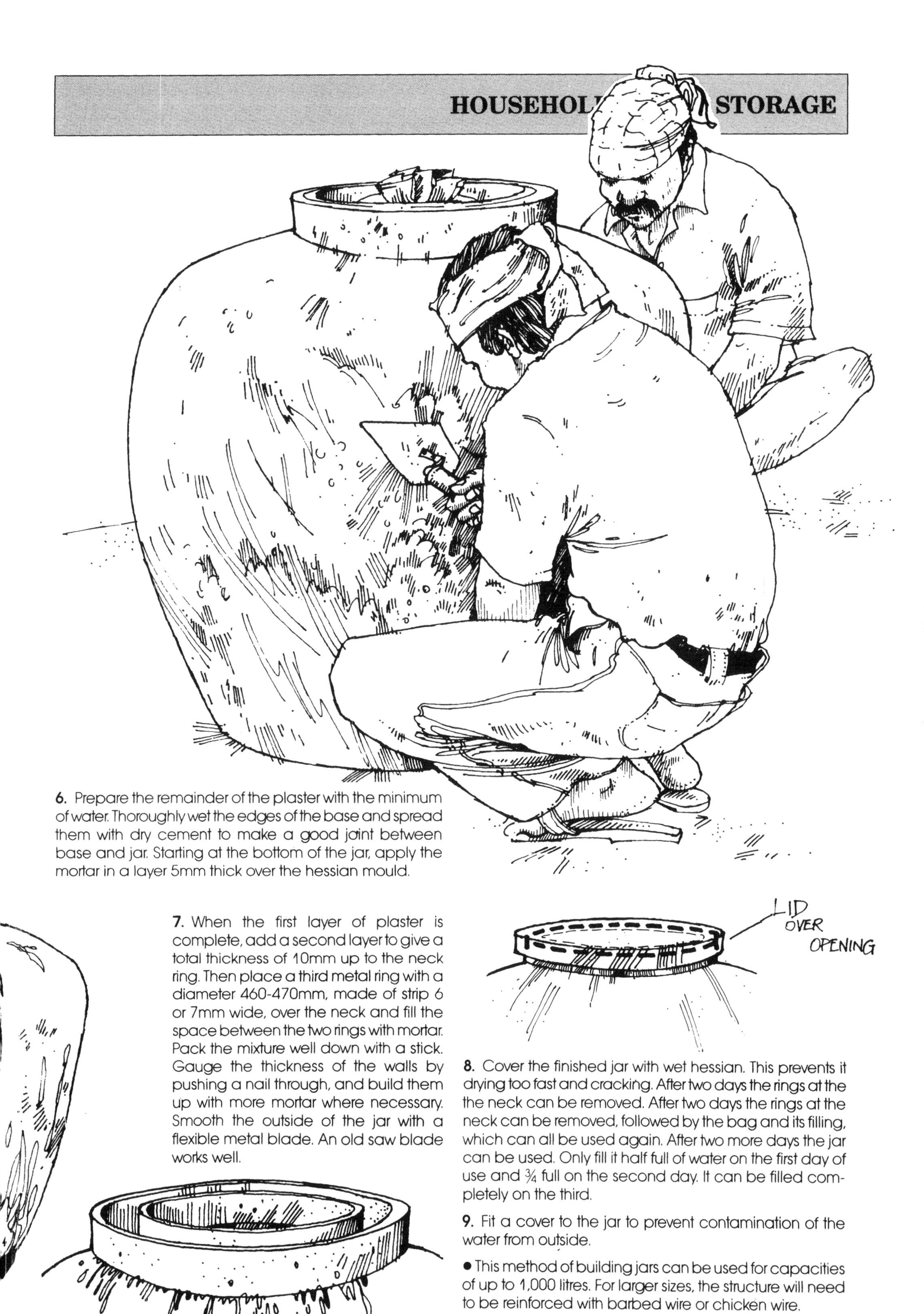

6. Prepare the remainder of the plaster with the minimum of water. Thoroughly wet the edges of the base and spread them with dry cement to make a good joint between base and jar. Starting at the bottom of the jar, apply the mortar in a layer 5mm thick over the hessian mould.

7. When the first layer of plaster is complete, add a second layer to give a total thickness of 10mm up to the neck ring. Then place a third metal ring with a diameter 460-470mm, made of strip 6 or 7mm wide, over the neck and fill the space between the two rings with mortar. Pack the mixture well down with a stick. Gauge the thickness of the walls by pushing a nail through, and build them up with more mortar where necessary. Smooth the outside of the jar with a flexible metal blade. An old saw blade works well.

8. Cover the finished jar with wet hessian. This prevents it drying too fast and cracking. After two days the rings at the the neck can be removed. After two days the rings at the neck can be removed, followed by the bag and its filling, which can all be used again. After two more days the jar can be used. Only fill it half full of water on the first day of use and ¾ full on the second day. It can be filled completely on the third.

9. Fit a cover to the jar to prevent contamination of the water from outside.

- This method of building jars can be used for capacities of up to 1,000 litres. For larger sizes, the structure will need to be reinforced with barbed wire or chicken wire.

Improving water quality by filtration

River water may be the only source available at times, but it is often dirty and not hygienically safe. A simple treatment system to supply up to eight families can be built using four covered jars, four valves, plastic tubing, sand (0.5-3mm and 0.5mm grade) and gravel, as follows.

The tubes can be fitted to jars either by making holes in the walls when the mortar is still soft or by cutting holes with a small hammer and a chisel (made from a sharpened screwdriver). Tubes can then be sealed in with cement mortar.

Jar 1 is for the storage and settlement of the untreated water and should be as large as possible. Valve A is fitted at the bottom and is used for cleaning out the jar. Valve B is fitted about 100mm above A and controls the flow of water to jar 2. Jar 1 is raised above the other jars in the system to provide pressure to push water through.

Jar 2 is an upward flow filter which will remove much of the coarse dirt. A 250 litre jar can treat 20 litres of water an hour. Back-wash the filter weekly by closing valve B and opening valve C to drain the jar. Clean, but not hygienically safe, water can be drawn from valve D.

Jar 3 is a downward flow filter. A 250 litre jar can treat 20 litres an hour.

It is cleaned by draining the standing water off the top and scraping the top 20mm of sand away every two or three months.

Jar 4 is the container for safe drinking water.

- Tests in Thailand show that the treated water can be free of faecal bacteria for prolonged periods of use.

- Information supplied by S. B. Watt (UK), H. Mann (UK), Mr Opas (Siam Cement Co, Bangkok, Thailand), Professor R. Suwanik (Siriraj Hospital, Bangkok, Thailand).

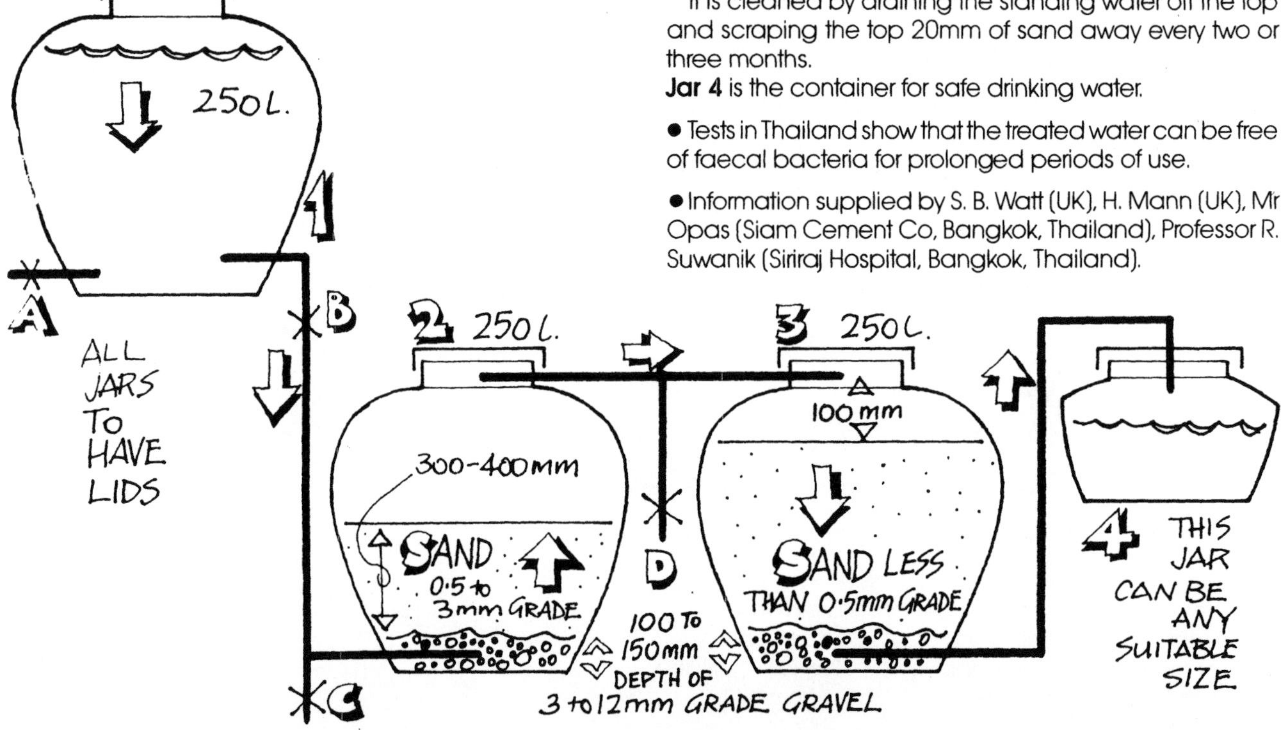

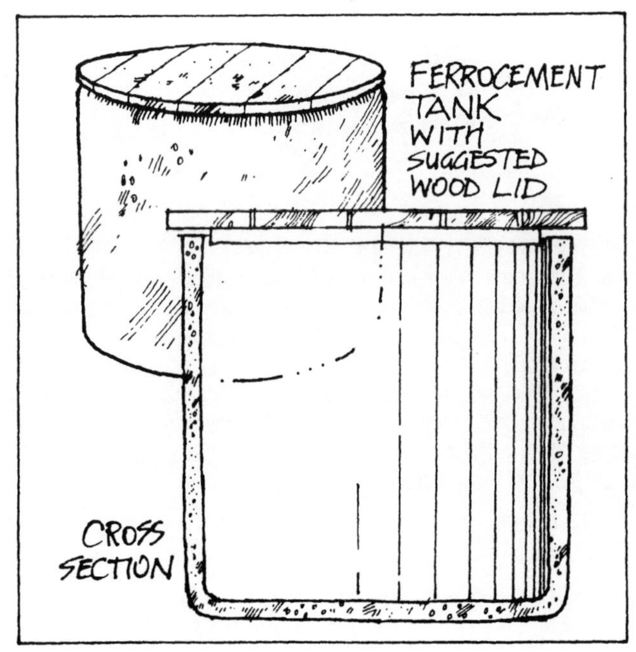

References

1. GRET, *Dossier No 4: the construction of cisterns, collection and storage of rainwater.* GRET, 30 Rue de Charonne, 75011 Paris, France.
2. *Water for the world technical notes.* Institute of Rural Water, Agency for International Development, Washington DC 20523, USA.
3. Nissen-Petersen, E. *Rain catchment and water supply in rural Africa: a manual*, Hodder and Stoughton, UK, 1982, 96pp.
4. Watt, S. B. *Ferrocement water tanks and their construction*, IT Publications, London, 1978, 118pp.
5. Mann, H.T. and Williamson, D. *Water treatment and sanitation*, IT Publications, London, third revised edition 1982, 96pp.

2. An introduction to pit latrines

Pit latrines

- can be as healthy as waterborne sewerage
- can be built by the users
- use a minimum of imported materials
- are low cost
- can be easily maintained by the users

A pit latrine is a way to deposit excreta

directly

or indirectly

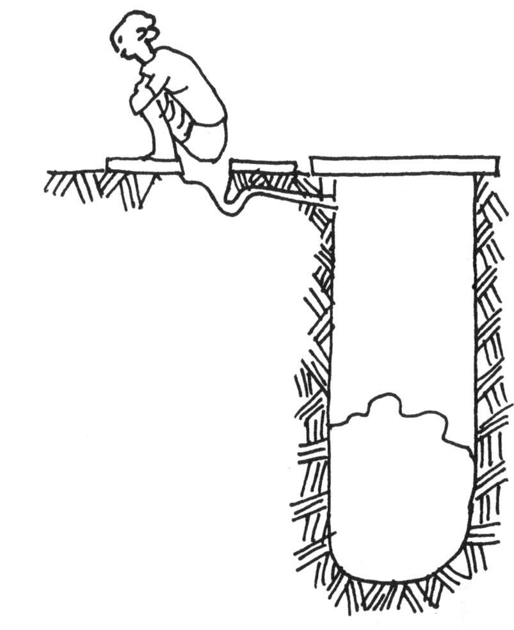

into a hole in the ground.

In the pit, excreta are decomposed into gases, liquids and solids.

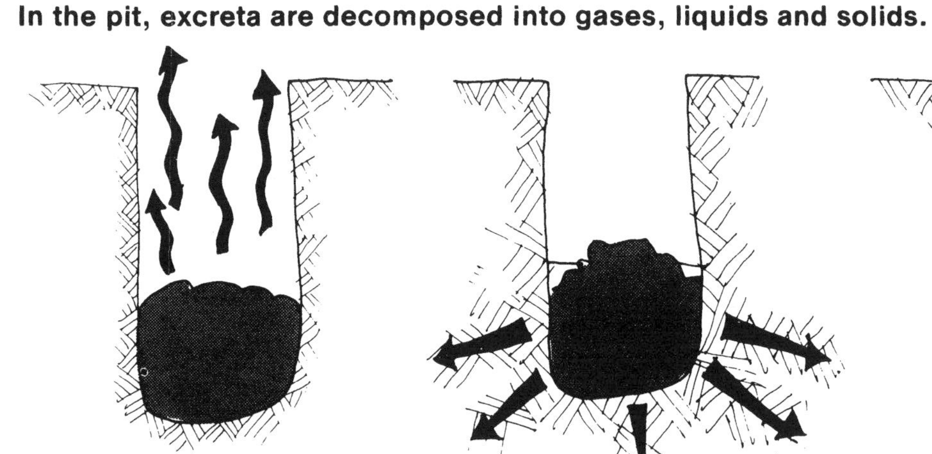

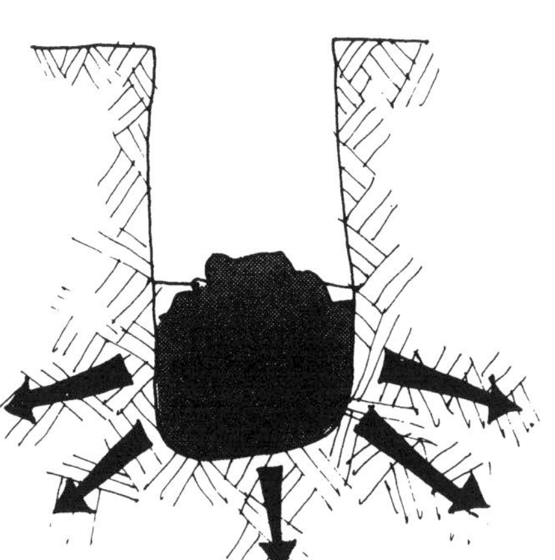

Gases escape to the atmosphere.

Liquids soak into the soil.

Solids which remain become harmless after a year and can be dug from the pit and used as fertilizer.

Cleanliness & health

Even a well-constructed latrine can spread disease unless it is kept clean.

The excreta of small children should be cleared up and put in a pit latrine until they can use the latrine properly themselves. Children's excreta are very infective.

Everyone should wash their hands thoroughly – with soap and a brush if possible – after using the latrine.

Unhealthy pit latrines

The pit is too shallow or too full, with the contents too close to the user, so that they smell bad and spread disease.

The slab over the pit does not allow water to drain away, so mosquitos can breed in it and the rough surface harbours hookworm larvae.

The pit sides are unsupported so that water can flow in, and they collapse.
The floor is made of untreated timber, so that it will collapse if the wood is eaten by termites.

The pit is open, so that flies can breed in the excreta.

The pit is wet, so that *Culex pipiens* mosquitos, which spread the disease filariasis, can breed. Filariasis sometimes develops into the condition known as elephantiasis.

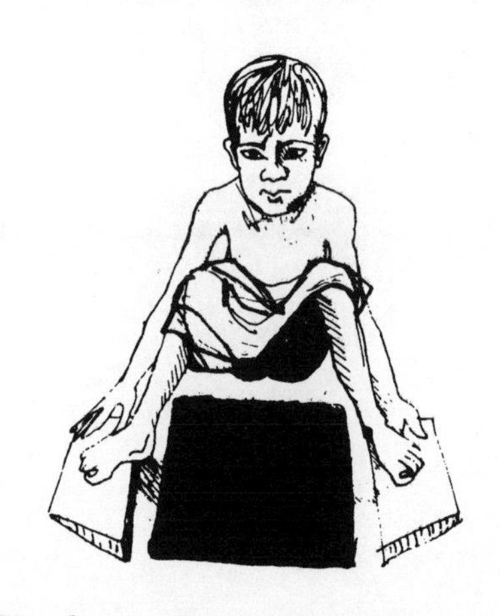

The foot-rests are too far apart and the hole in the slab is too large, so that the latrine is uncomfortable for children to use, and they are in danger of falling in the pit.

Healthy pit latrines

Large pits
Pits should ideally be at least 3m deep and 1-1.2m in diameter. Pits this size may be used for up to twenty years.

Lined pits
Linings may be of treated timber, natural stone, sandcrete blocks and bricks, concrete pipes, corrugated iron, or flattened oil drums.

The hole in the squatting slab should be large enough to prevent fouling but not so large that people fall down the hole. It may be rectangular, square, oblong, oval or circular.

In some places people use a mixture of cow-dung and mud to make a smooth surface for the slab.

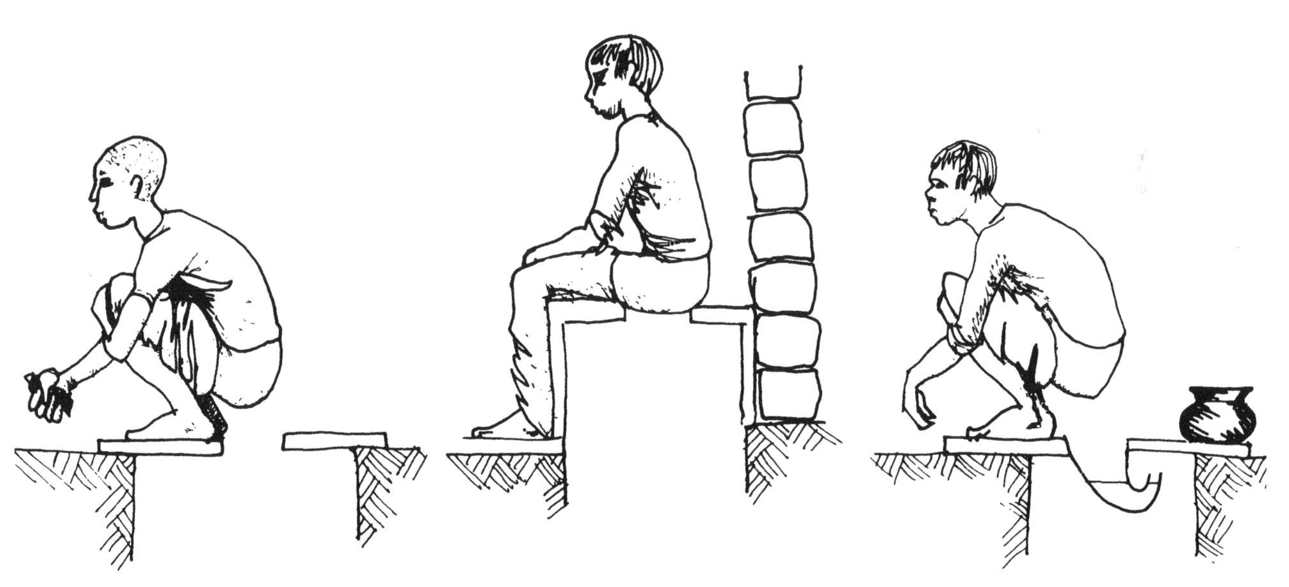

Where people squat to defaecate, and use paper, leaves, stones and other hard material to clean themselves afterwards, a pit usually has a reinforced concrete or ferrocement squatting slab over the hole.

Seats can be built for people who prefer to sit.

Where people clean themselves with water, a water seal to prevent smells rising from the pit can often be installed below the slab.

Varieties of pit latrines

Simple pits. These can be smelly and fly-ridden, especially if they are too shallow. They should ideally be 3m deep and 1-1.2m in diameter.

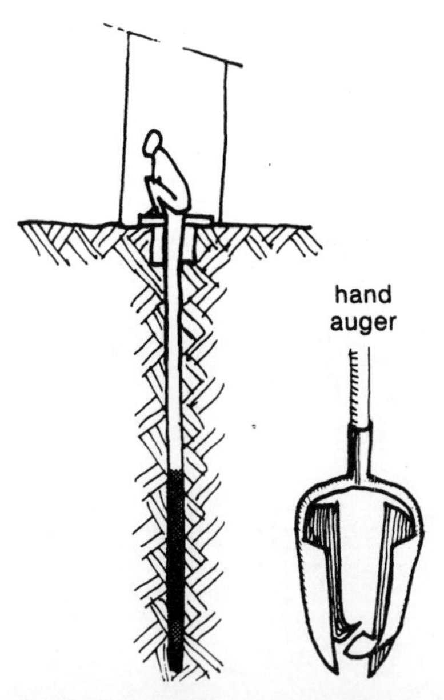

Borehole latrines. These can be quickly constructed with a hand augur. They are especially useful where a quick sanitation solution is essential (eg. refugee camps) and where people are likely to move around regularly. However, the small diameter of the hole means that it is likely to foul, block and fill quickly.

Ventilated pit latrines. These have been designed to reduce smell by drawing odour away from the squatting slab and to discourage flies and mosquitos.

- **ZIMVIP** (Zimbabwe Ventilated Improved Pit Latrine). Its main innovation is a superstructure with a spiral cross-section.

Mound or step latrine. This type is suitable for areas where you cannot dig into the ground, eg. where the water table is high or there is hard rock near the surface. However, the user is close to the excreta.

Offset latrine. The waterseal and seat can be on firm ground. The pit can be in softer, easier-to-dig ground, and can be protected to prevent collapse. The slab can be small, as it does not have to act as a lid for the pit, useful where cement and other reinforcement are difficult to get.

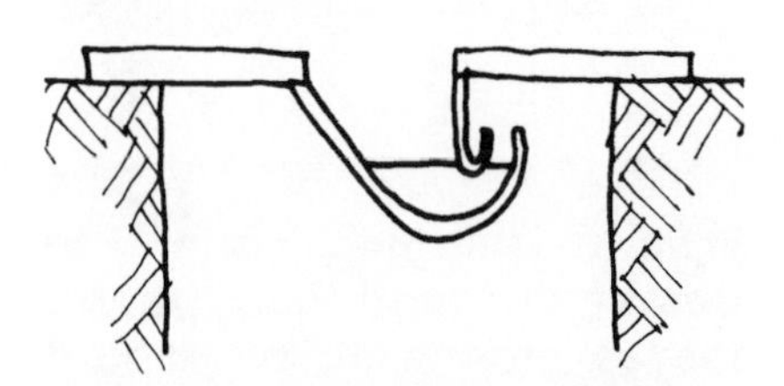

Pour-flush latrine. These are most suitable where people use water for anal cleansing. Unfortunately, the pan is easily blocked and tends to get broken when users attempt to clear it with a stick.

Of course, many pit latrines are combinations of more than one of these options.

3. Protecting a spring

There are two main types of springs

Gravity springs occur where groundwater emerges at the surface because an impervious layer prevents it seeping downwards. This type usually occurs on sloping ground and its flow changes with variations in the height of the water table.

Artesian springs occur where groundwater emerges at the surface after confinement between two impervious layers of rock. The flow is very nearly constant.

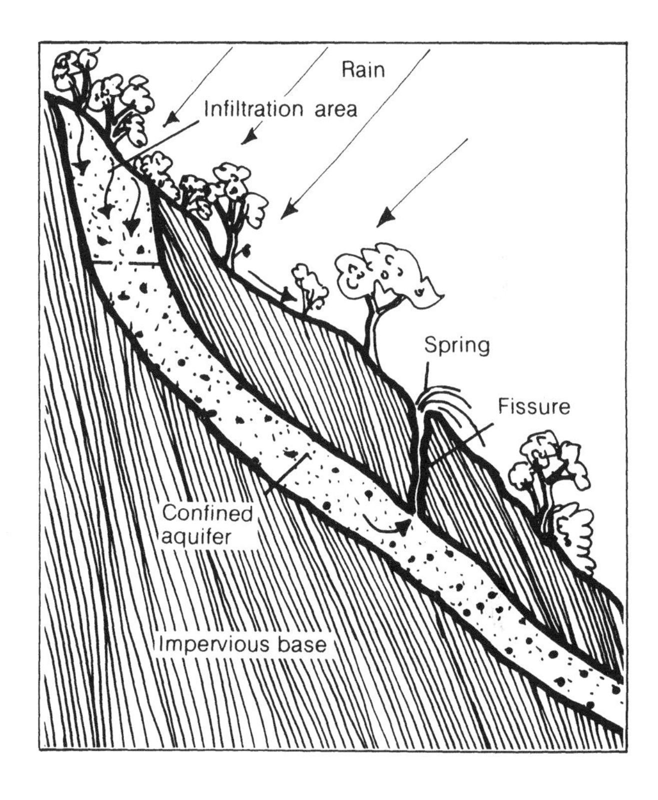

Selecting a spring

- Consult local people, who will know where the springs are in the area. They will also be able to tell you which ones stop flowing during the dry season. The best time to measure the flow is in the dry season.
- Ask the villagers if the flow is ever greater than when you measure it. If the maximum flow is greater than 10 litres/second, a simple spring box will not be strong enough.
- Investigate around the source to make sure the 'spring' is not really a stream which has gone underground and is re-emerging.
- If the spring is to be connected to a piped water system, it should be well above the village and as inaccessible as possible to minimize pollution by children and live-stock.
- If people are to collect water directly from the spring it should still be on higher ground than the village it serves, but as nearby as possible.
- Intakes should not be built on swampy ground.
- Intakes should be located where they will not be threatened by land erosion or flood waters or where the water will cause erosion.
- Try to incorporate other facilities requested by villagers, eg. for washing clothes or watering animals.

You will also need the following information:

- Check whether the water looks clear
- The flow of the spring (litres/second)
- Possible sources of contamination
- Details of water rights in the area

PROTECTING A SPRING

Each spring is, of course, a special case, but building an intake box is one method of protecting a spring from contamination.

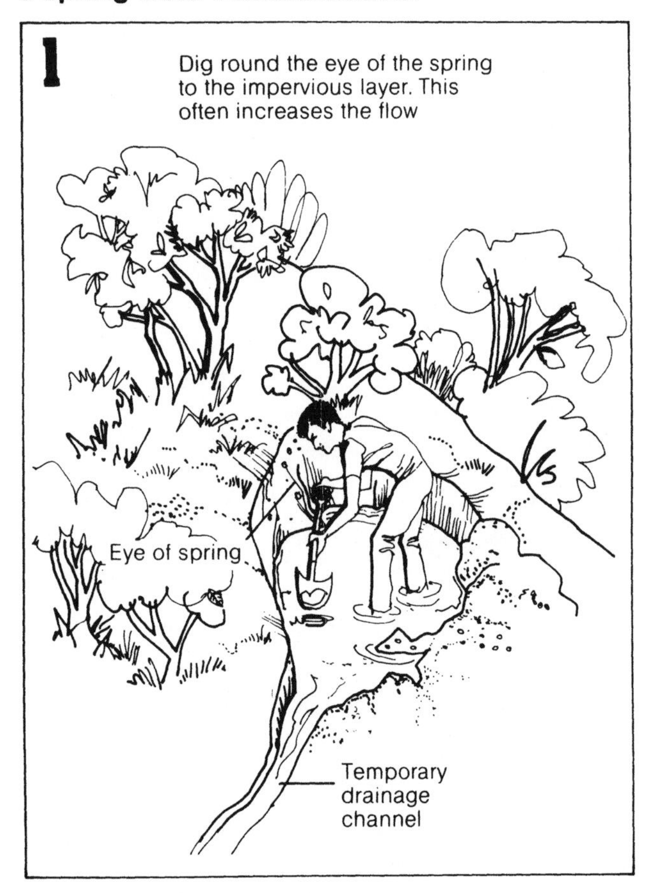

5
Front and side walls from locally available materials
eg. stone masonry
cement
brick
reinforced concrete
Wash-out pipe
2in or 3in diameter GI pipe
Clean spring box at least annually

6
Outlet pipe
If to be connected to a piped system, a gate valve will be needed downstream
Galvanized iron (GI) outlet pipe
If people will collect water direct from the source, use 2½-3in pipe for flows of 1-2 litres/sec. If piped system, size of pipe is determined by designed pipe line. See reference 1

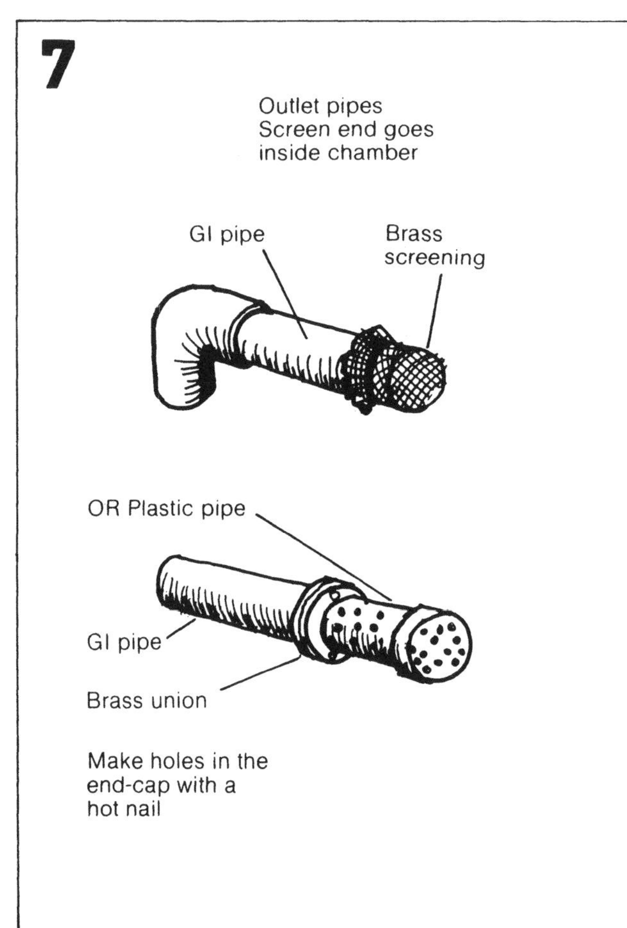
7
Outlet pipes
Screen end goes inside chamber
GI pipe
Brass screening
OR Plastic pipe
GI pipe
Brass union
Make holes in the end-cap with a hot nail

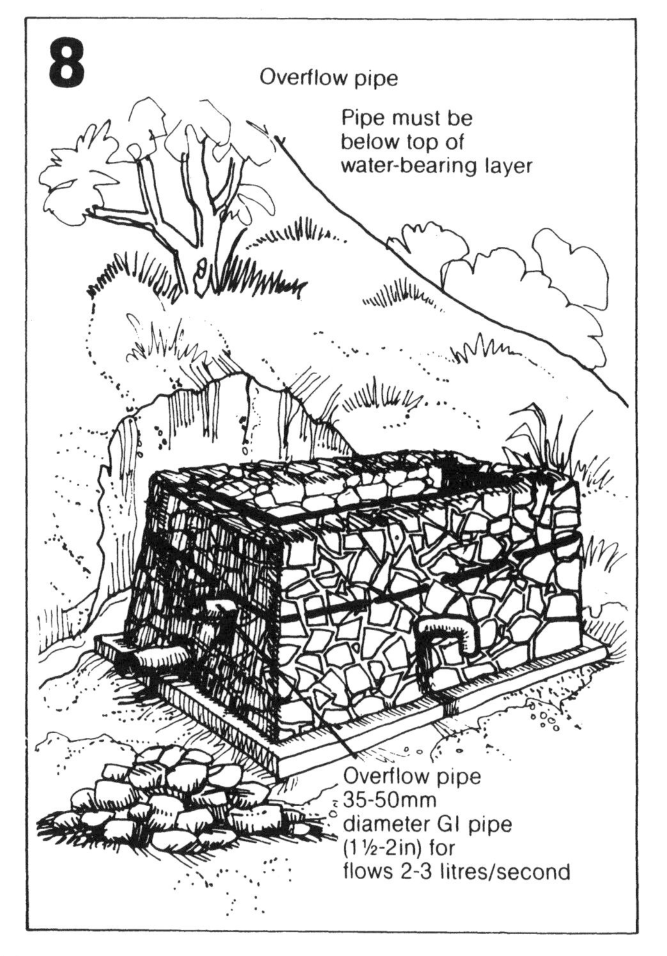
8
Overflow pipe
Pipe must be below top of water-bearing layer
Overflow pipe
35-50mm diameter GI pipe (1½-2in) for flows 2-3 litres/second

PROTECTING A SPRING

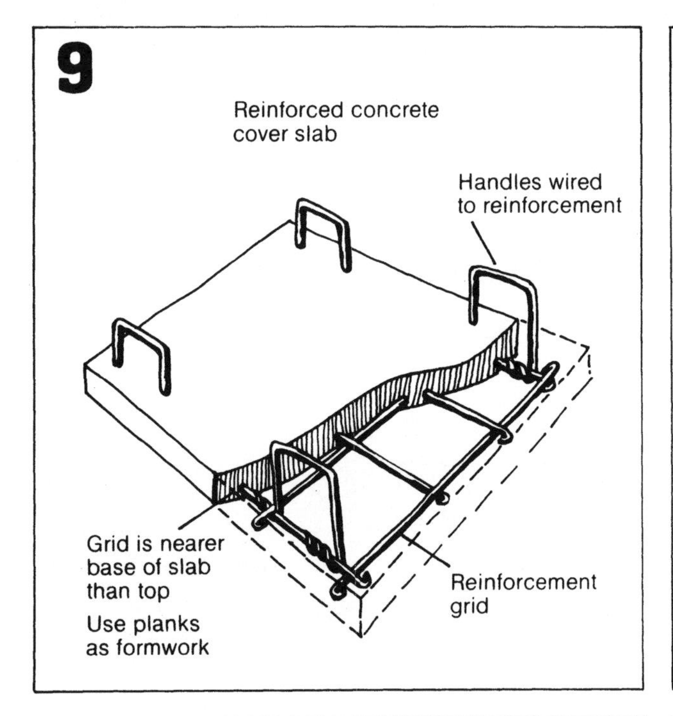

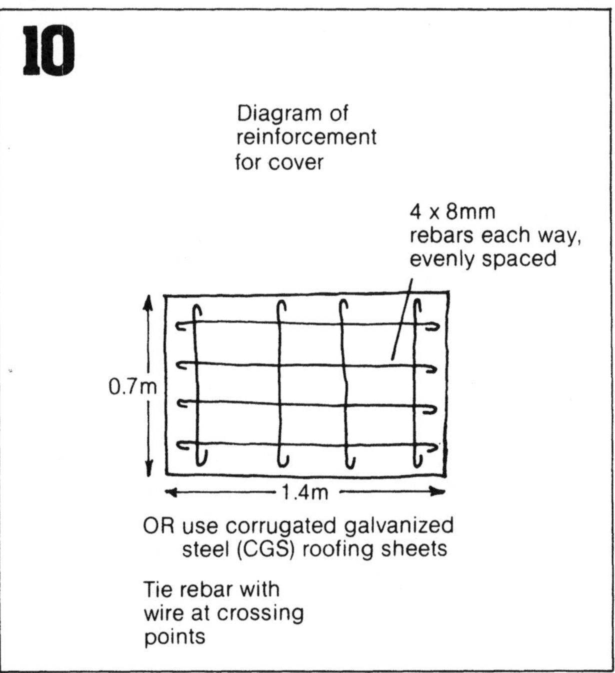

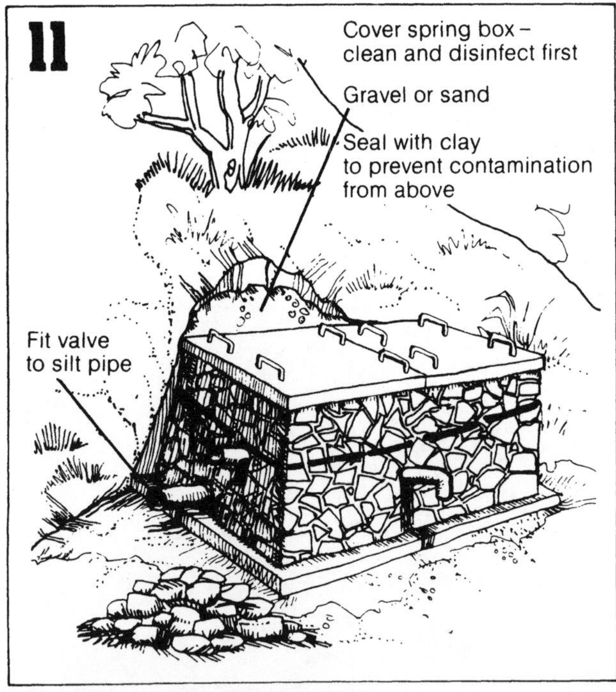

For more information:

1. **Jordan, T. D. Jnr. *A handbook of gravity-flow water systems*, Intermediate Technology Publications, London, UK, 1984.**
2. **Cairncross, Sandy, and Feachem, Richard. *Small water supplies*, Ross Bulletin No10. The Ross Institute, London, UK, January 1978.**
3. **Wagner, E. G. and Lanoix, J. N. *Water supply for rural areas and small communities*, Monograph No42, WHO, Geneva, 1959.**

Many thanks to Vanessa Tobin and Sandy Cairncross of the Department of Tropical Hygiene at the London School of Hygiene and Tropical Medicine.

4. Lining a hand-dug well

There are as many different ways of building a well as there are different sites.

Alternative methods

1. Sinking caissons (Concrete rings).

2. Reinforced concrete or ferrocement cast *in situ* above waterline, concrete rings sunk below waterline.

3. Masonry lining of burnt bricks above waterline, caisson made of blocks with cutting ring below waterline.

4. Galvanized iron rings bolted together as a temporary measure for emergencies.

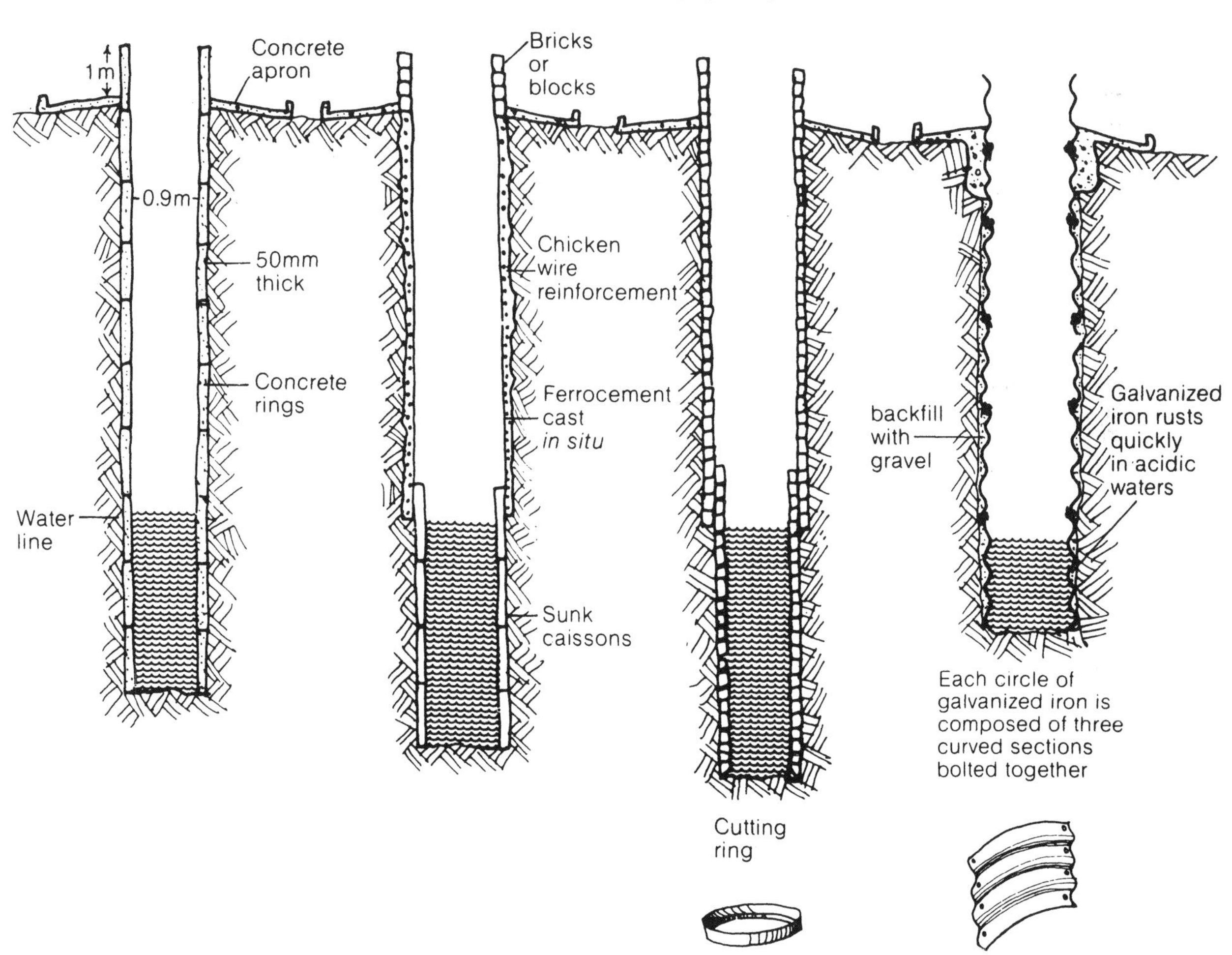

This Technical Brief concentrates on sinking caissons as one of the most useful methods. It is safe, efficient and economical where the cost of equipment, notably the steel moulds for casting the concrete rings, can be spread over several wells. The skill of the well-diggers will also build up with experience. A tube of rings is built upwards and is allowed to sink under its own weight to its final position as the soil is excavated from within it.

LINING A HAND-DUG WELL

Before you start digging —

Know the community, meet leaders, talk to the women who will be using the well. Involve the community, which should provide the money for materials, and provide skilled and unskilled labour.

Look at existing water sources. If you are going to convert an existing water hole, it should be well away from pit latrines but convenient to reach.

Is a well the simplest solution?

Check dry and wet season water levels.

Concrete rings

These need to be cast from steel moulds which must be accurately made. Rings can either be cast beside the well or be transported to the site. Concrete should be made in the proportions 1:2:4 cement:sand:aggregate (if materials are of good quality), well tamped into the mould and kept covered and damp for at least five days to cure.

Shorter rings can be lifted more easily, but need to be thicker. If reinforcement is omitted, the walls must also be thicker.

Finished ring

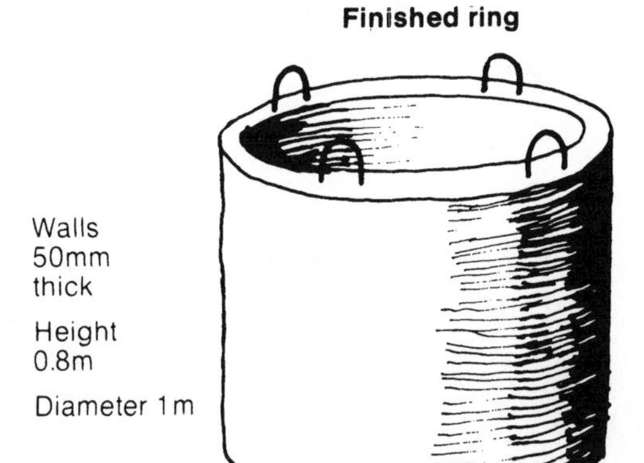

Diagram of reinforcement

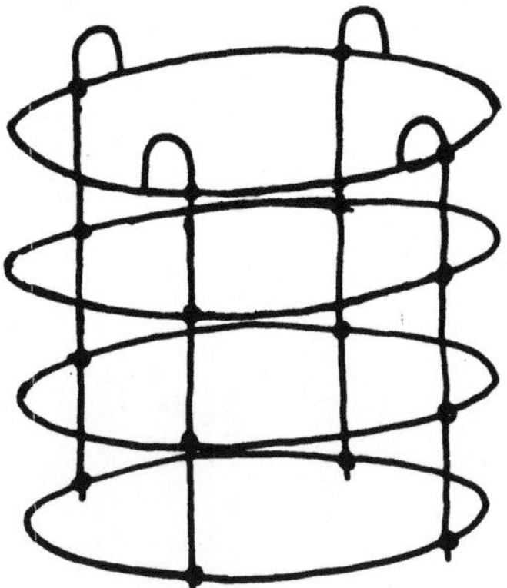

Quantities for 1 ring (1:2:4 ratio): Cement – 1 bag. Clean sand – 45 litres. Gravel – 90 litres. not larger than 15mm diameter. Rebar – 16m, of 6mm diameter.

The water hole

Clean out mud and organic matter. Dig the hole wide enough to accommodate the ring and as deep as it is safe to go without the sides collapsing.

LINING A HAND-DUG WELL

Sinking rings

Suspend the first ring on ropes as shown, using about 15 people to hold the ropes to be safe, and lower it until it sits *level* at the bottom of the hole.

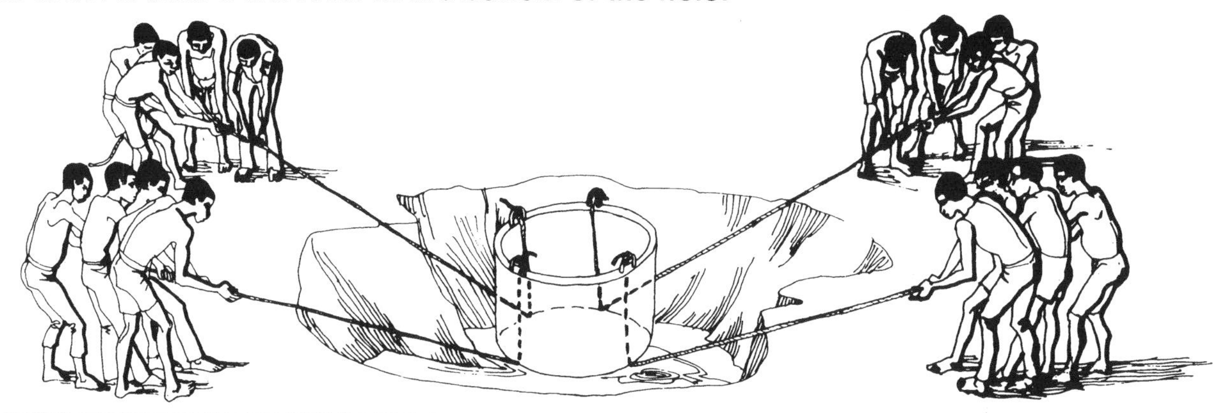

Continue digging inside the ring.
Wells can be de-watered by bucket and hard work, or by a small diesel, petrol or hand-operated diaphragm pump.

Lower further rings as the first one sinks. The deeper the well the larger the amount of storage and in some cases infiltration. The well is used during the day and refills at night.

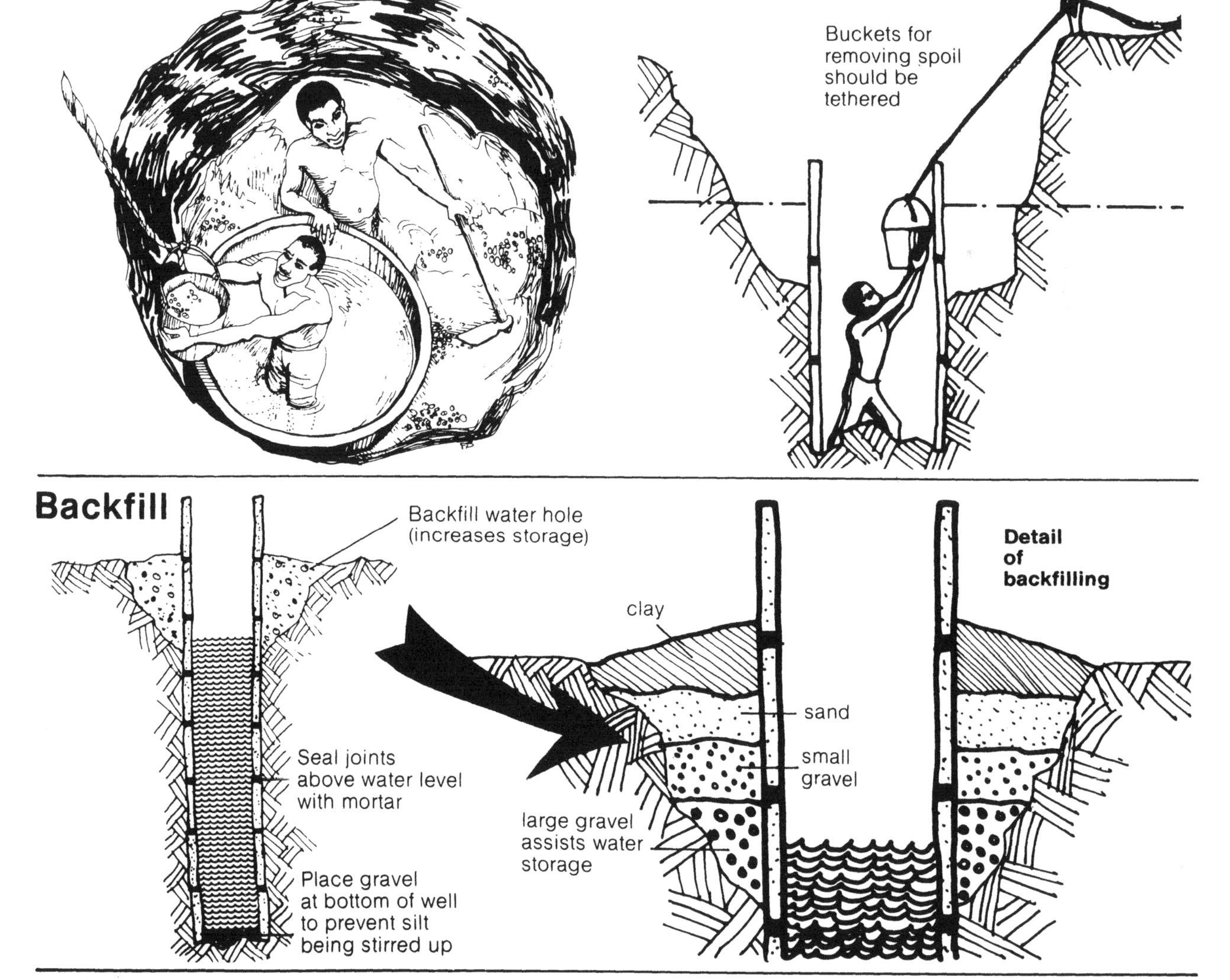

Backfill

Safety

Workers should wear helmets and buckets should be tethered while they are down the hole. Keep engine-driven pumps away from the hole to avoid the build-up of toxic fumes.

But don't stop there . . .

Your well can now provide fresh water, as long as it is properly finished.

The surrounds soon get dirty, allowing contaminated water to seep back into the well. Puddles of spilt water transmit hookworm through feet.

So disinfect the well with chlorine solution or bleaching powder. Mound up the earth around the rings so that water drains away from the well, and compact the earth firmly.

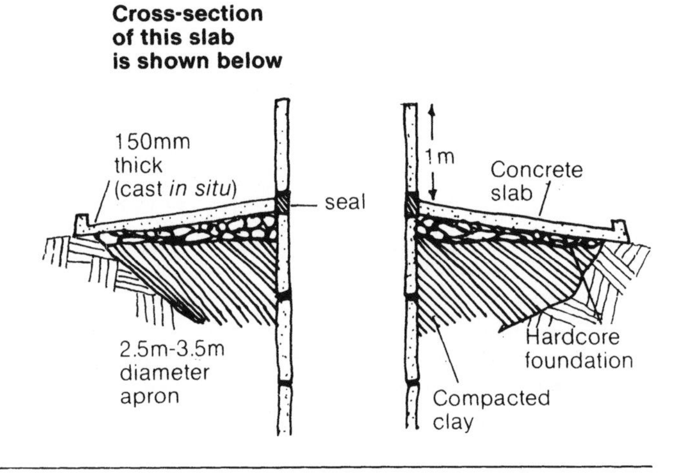

Lay hardcore, then cast a concrete apron – reinforced if possible. Make raised edges and cast a channel for drainage.

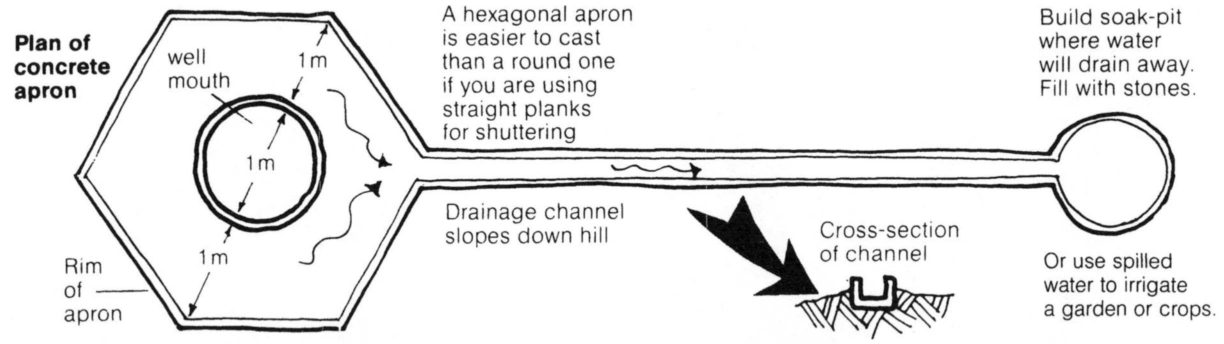

Pump vs bucket

If handpumps are already being used in the area and can be maintained, cast a slab to fit over the well to mount the pump. (Handpumps designed for bored wells can often be adapted.)

Simplest solution is often a pulley or windlass and a bucket which stays at the well.

Make sure everybody realizes how important it is to keep the well and its surrounds clean. Plant flowers and shrubs. It is advisable to fence the site.

The detail shows a tripod made of old rising mains with the ends flattened, bolted together, fitted with a pulley.

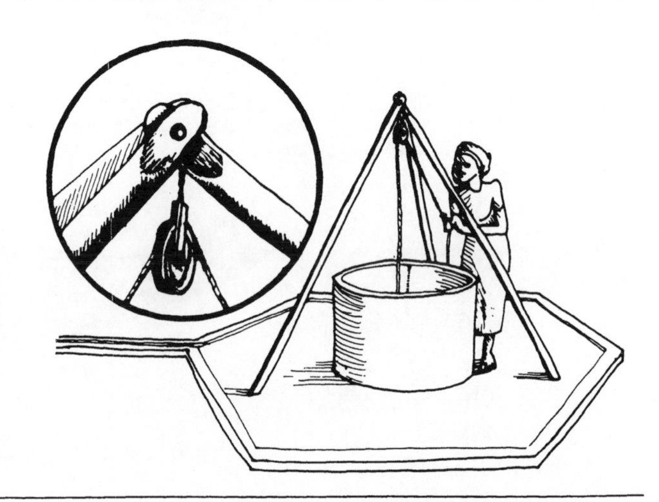

For more information

1. Watt, S. B. and Wood, W. E. *Hand dug wells and their construction*. Intermediate Technology Publications, London, UK, 1976.
2. DHV Consulting Engineers. *Shallow wells*, DHV Consulting Engineers, PO Box 85, Amersfoort, The Netherlands, 1978.

Compiled by Valerie Curtis. Illustrations by Frances Stuart.

5. Slotted bamboo tubewell screen

Where the ground is soft enough, tubewells can be constructed by hand with simple drilling tools. They can be much easier to install and cost much less than large diameter wells.

The storage capacity of tubewells is limited because of their small diameter. Therefore the rate at which water flows into the well from the surrounding soil is critical. This is governed mainly by the design and positioning of the well screen and how the well is developed.

The term 'well development' refers to the process of removing the finer particles from the aquifer immediately around the well screen and rearranging those remaining so that:

- Fine particles cannot subsequently be washed out so that they clog the openings of the screen or damage the pump,
- Water can flow more easily into the well because the permeability of the earth immediately surrounding the screen is improved.

Cross section of a hand-drilled well

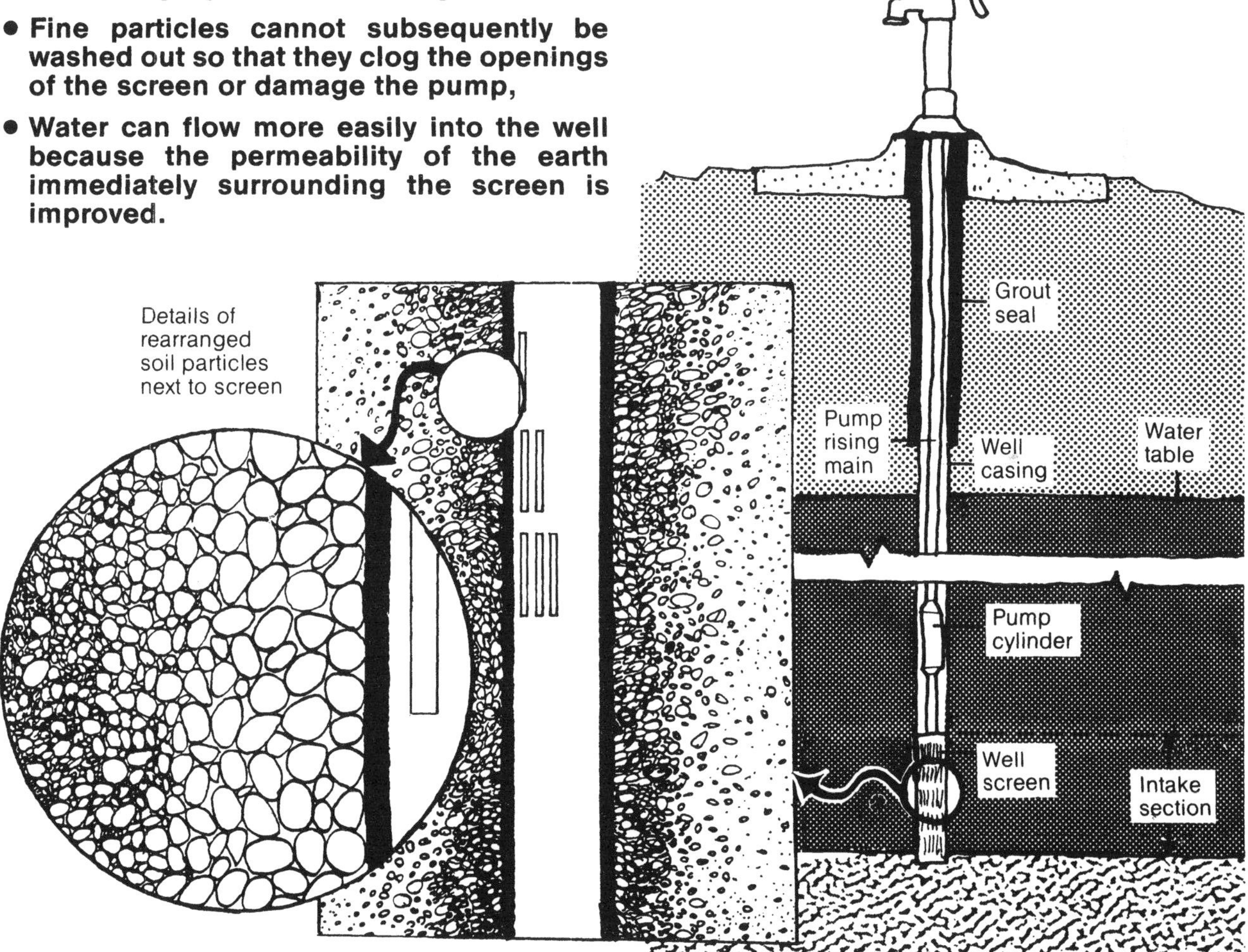

A good screen should hold back sand particles after the initial development of the well and let water flow in freely at the same time. It should be strong enough to withstand handling during installation. It should also be resistant to corrosion.

This Technical Brief concentrates on one option for constructing well screens: bamboo. In north-east India and Bangladesh, many thousands of bamboo tubewells have been installed, and some have performed satisfactorily for seven years or more. Even if they last for a shorter time, their very low cost makes replacement feasible. They are an alternative to commercially-produced well screens made from brass, steel or plastics.

The technique described here can also be used to slot *pvc* pipe.

SLOTTED BAMBOO TUBEWELL SCREEN

The slotting machine

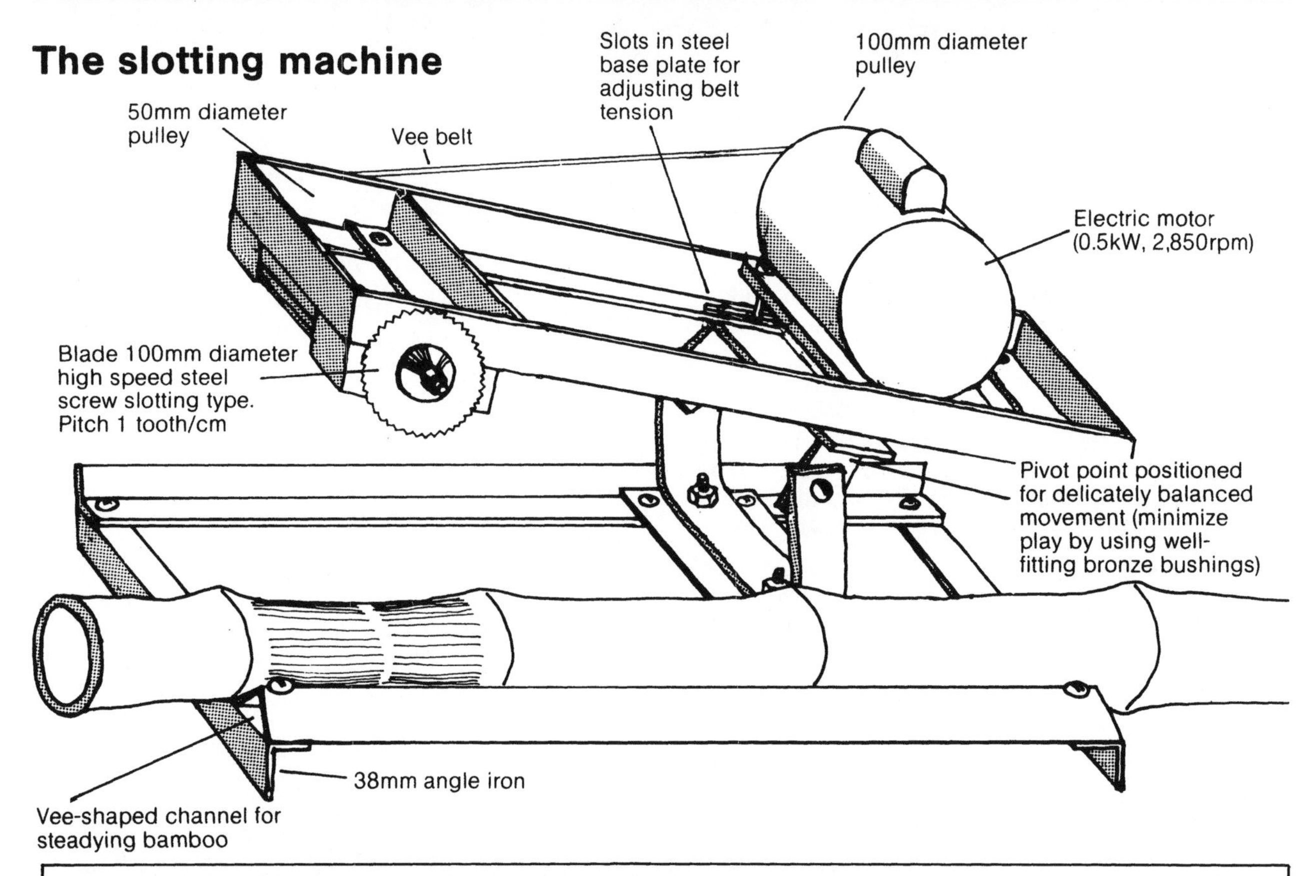

Note: For clarity, this drawing shows the machine without safety guards on belt and blade. These are essential for safe operation.

Cutting the slots

A 0.5kW motor will drive a 100mm diameter blade through dry bamboo.

The power requirement increases dramatically if the bamboo is wet, or if multiple blades are used to cut more than one slot at a time.

It may be easier to hold the bamboo in its vee-shaped holder than to clamp it in, due to its uneven shape and size.

Slots should be a maximum of 100mm long.

If it is necessary to move the bamboo longitudinally during cutting, beware of the blade binding and making an uneven slot.

Rotate the bamboo by hand before starting the next parallel slot.

A minimum of 6mm should be left before starting the next series of slots. This will ensure that the screen is strong enough.

The slots should ideally take up more than 10 per cent of the screen area

If the slot is narrow enough, there will be no need to gravel-pack the well. Carefully-graded gravel is placed round the screen during construction of the tubewell to assist in well-development.

For best results, bamboo slotted when dry should be soaked in water before installing, as slots shrink by about 30 per cent on immersion in water.

Do not leave bamboo in sunlight and open air, where it will decay quickly unless it has been treated with preservative. It has a long lifespan providing it is permanently saturated, however.

Removing the nodes

Bore the nodes of the bamboo out manually using a home-made bit attached to 12.5mm diameter galvanized iron (GI) pipe or another length of bamboo (as shown in the top diagram). The bit can be arrowhead-shaped or star-shaped, consisting of four blades.

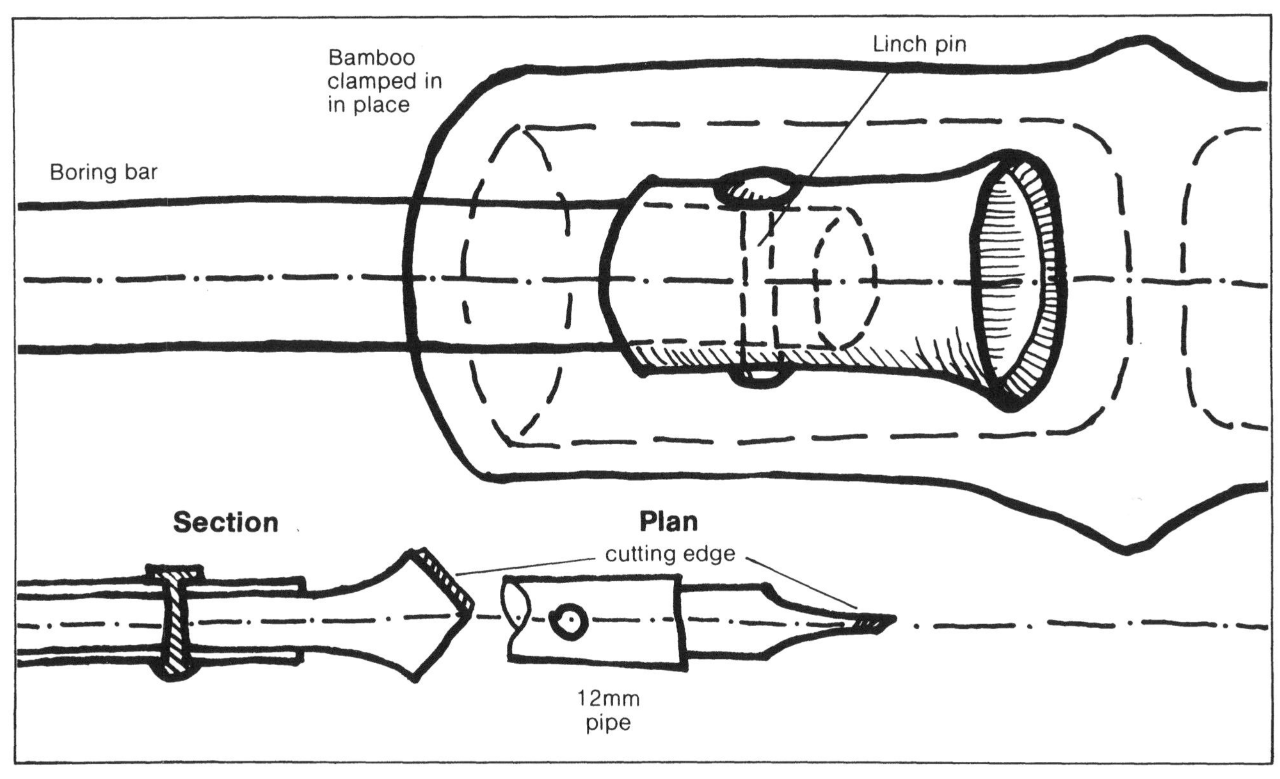

To make the bit illustrated (top), bell out the end of a short length of steel pipe to increase its diameter, and sharpen the edge. Slide the 12.5mm diameter steel pipe inside as a boring bar. Drill a small hole through the assembly and drive a nail through it as a linchpin.

Making joints

Bamboo to bamboo: use a bamboo socket caulked with cotton wool and tar, and nails.

Bamboo to GI: cut threads in the end of the bamboo filter and use a threaded GI socket.

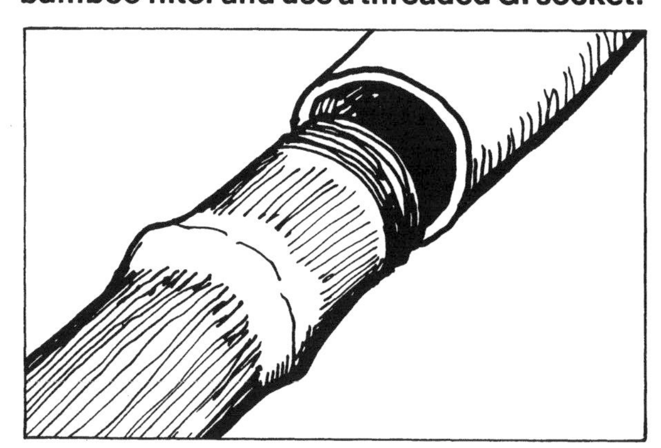

Sample installation

A 2.1m length of 75mm diameter screen with slots 100mm long and 0.9mm wide was installed in a 100mm diameter hole drilled to a depth of 12m. The rising main had a diameter of 38mm. The screen was gravel-packed with 10mm thickness of 3mm size gravel, and the well was developed by powerful pumping and back-washing. It yielded 200 litres/minute of water when connected to a 38mm centrifugal pump. The screen cost a tenth of the price of the cheapest available commercially manufactured screen.

Alternative: a bamboo and coir string screen

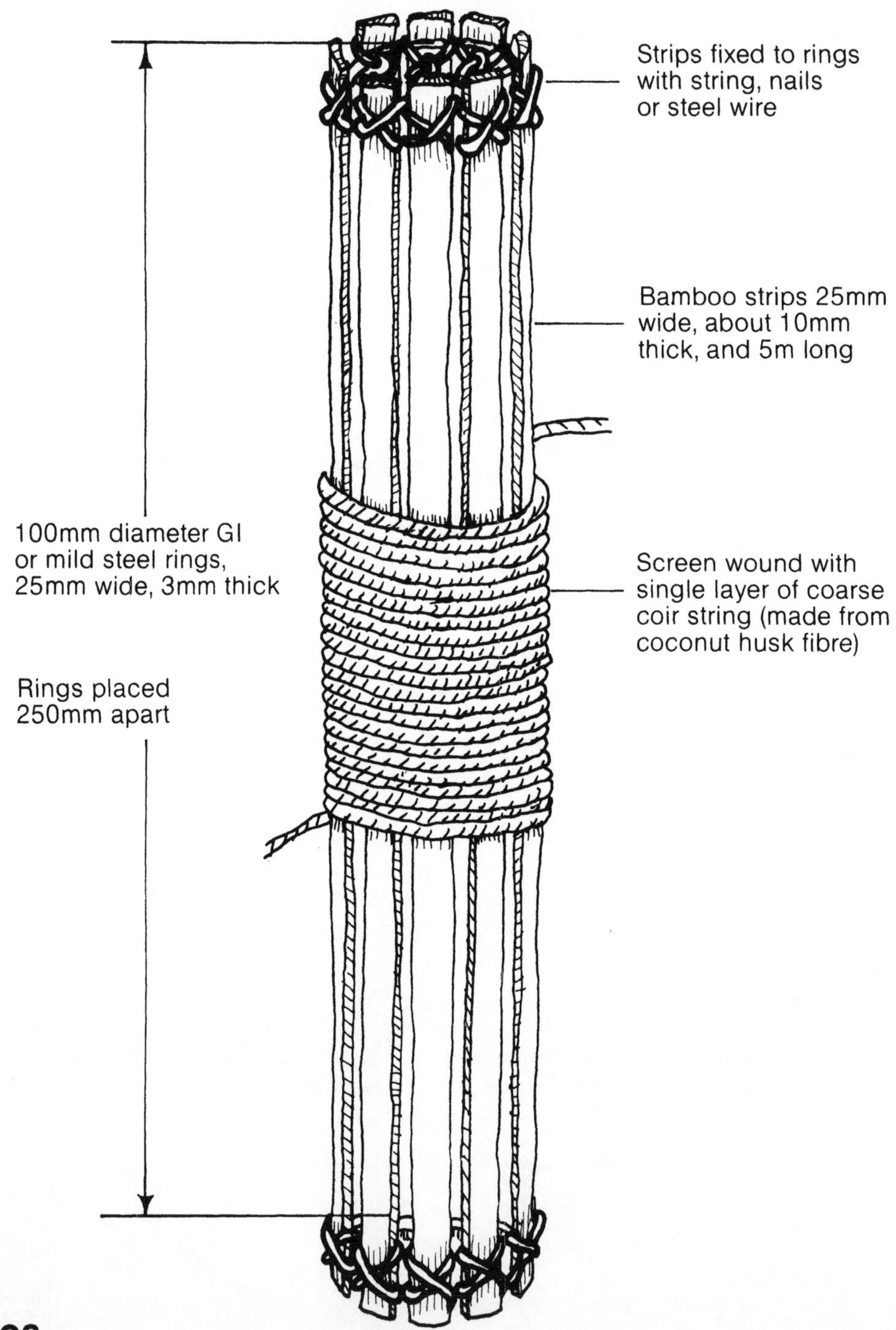

References

1. **Eaton, Bruce. 'Longitudinally slotted bamboo tubewell filters', *ADAB News*, India, August 1976.**
2. ***Using water resources*, Volunteers in Technical Assistance, Arlington, Virginia, USA, 1978.**
3. **Allison, Stephen V., Sternberg, Yaron M., and Knight, Robert. 'Well casings and screens from single stalks of bamboo, and a manually operated slotter', *Appropriate Technology* Volume 5 Number 1, ITDG, London, May 1978.**

Many thanks to Bruce Eaton.

6. Choosing a water-seal latrine

In parts of the world where people clean themselves with water after excreting, a latrine with a water seal pan can be used. The advantage of this over a latrine without a water seal is that the user is not in such direct contact with the latrine pit's contents. But construction is more costly and complex than a simple latrine, and a reliable source of water is needed to flush it.

This Technical Brief is designed to help you select the best type of water seal latrine for your local conditions. The right type of latrine has to be judged for each specific site: there can be no standard design for a whole country or even an area of a country. There are three types of water seal latrine, and in two of them the pit is offset from the latrine hut:

Direct

The pan is directly above the pit.
Advantage: The cheapest type of water seal latrine, needing the least amount of water for flushing.
Disadvantage: When the pit is full, the user has to build a new latrine or dig out the pit while the excreta at the top is still fresh. However, digging out the latrine is not too difficult if the above-ground structure is light-weight. People fear falling into the pit if the latrine is directly above it.

Offset

The pan is not directly above the pit.
Advantages: When the pit is full, a new one can be dug next to it. The pan does not have to be removed, and can be plumbed into the new pit.
Disadvantage: It is more expensive than a direct pit and takes more water to flush solids down the sewer pipe.

Double

There are two offset pits connected to the pan by a manhole.
Advantage: when one shallow pit is full, the other is used.
Disadvantage: It is the most expensive and complicated to construct and needs strict supervision during construction. It is also much more complicated from the point of view of user education, because the user has to understand fully how the latrine works. It uses more water than the direct pit.

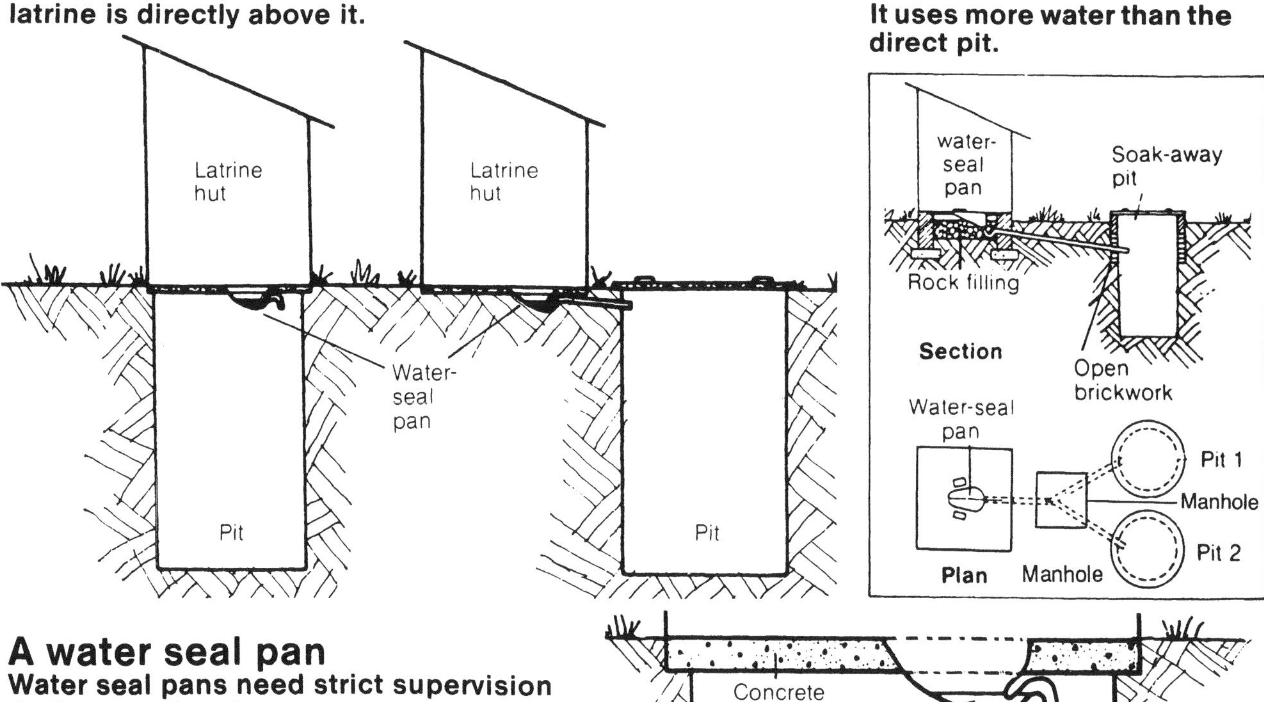

A water seal pan

Water seal pans need strict supervision during construction.

CHOOSING A WATER-SEAL LATRINE

Selecting the type of water seal

The key factor in selecting the best type of water seal latrine for each particular situation is the amount of water available to flush it.

Direct

Offset*

*A double pit has the same water requirement as an offset pit.

VIP

If 2-4 pints of water are not available for each flush, a water seal pan cannot be used. The alternative is a Ventilated Improved Pit Latrine (VIP) which has no water seal.

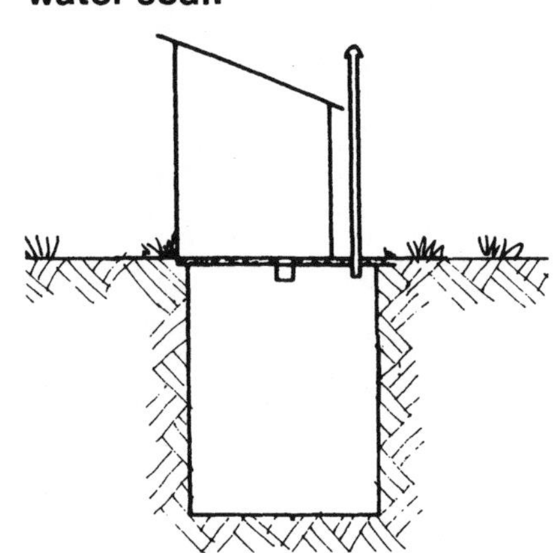

Selecting the type of pit

The deeper the pit, the longer it will last. It should ideally be at least 3m deep and 1-1.2m in diameter. Such a pit could be used by one family for 20 years.
You may not be able to dig as deep as 3m because

There is hard rock near the ground surface.

OR

The water table is too close to the ground surface.

In these cases,

The latrine can be built on a mound. The pit walls need to be built up at least 4ft (before the mound is constructed). A pit built like this might be expected to last about 8 years.

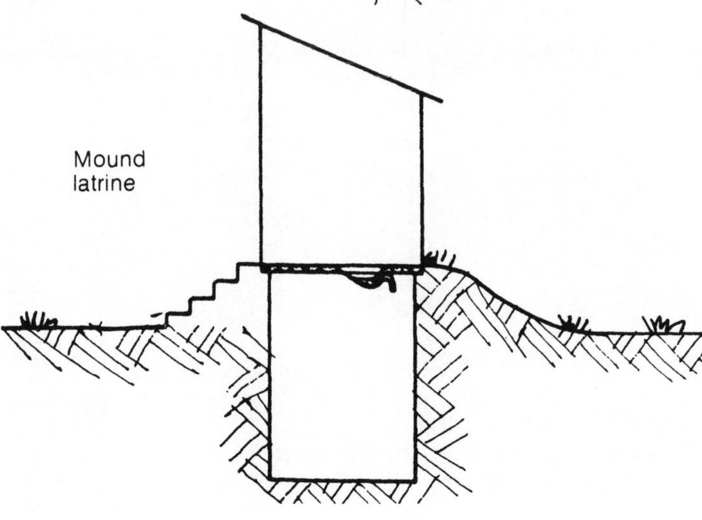

OR

A double pit could be used. It matters less than with a single pit if the pits are shallower than recommended, and fill up more quickly, because they can be used alternately. But one pit needs to be emptied every two years.

Construction

Dig the pit in the dry season when the water table is lowest.

If the soil is loose, all of the pit should be lined.

Even if the soil is very firm, the top 18 inches must be lined.

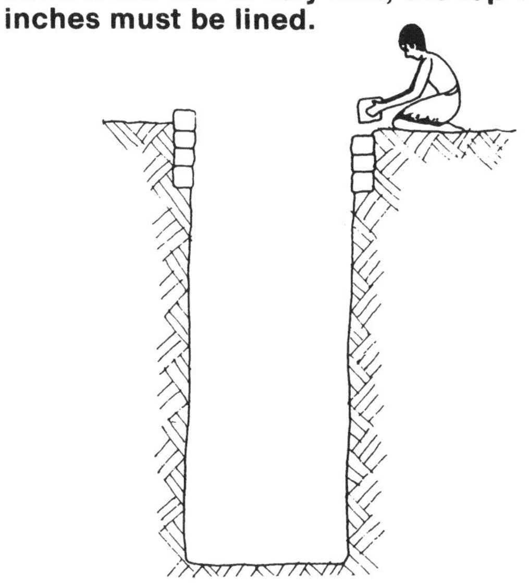

For the direct pit latrine, the slab and water seal pan should be securely fixed to the lining of the pit with cement mortar.

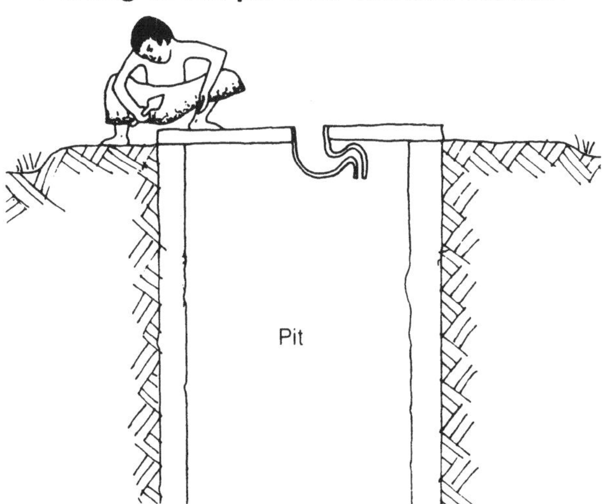

For the offset latrines, lay the pipe connecting the water seal pan to the pit.

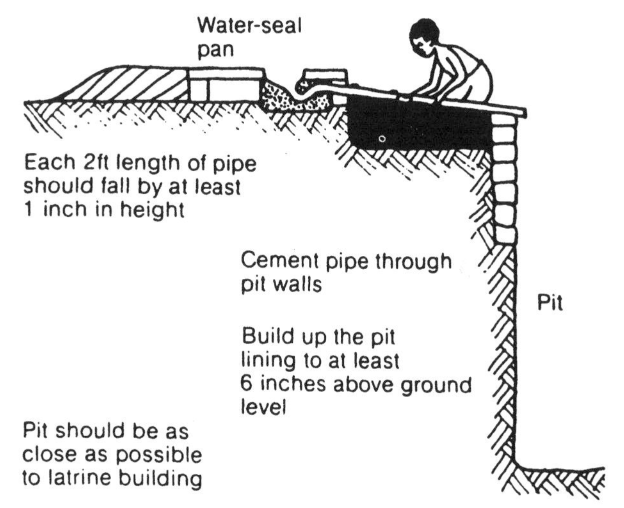

Cover the pit with a reinforced concrete slab. If the pit is large, two slabs may be needed.

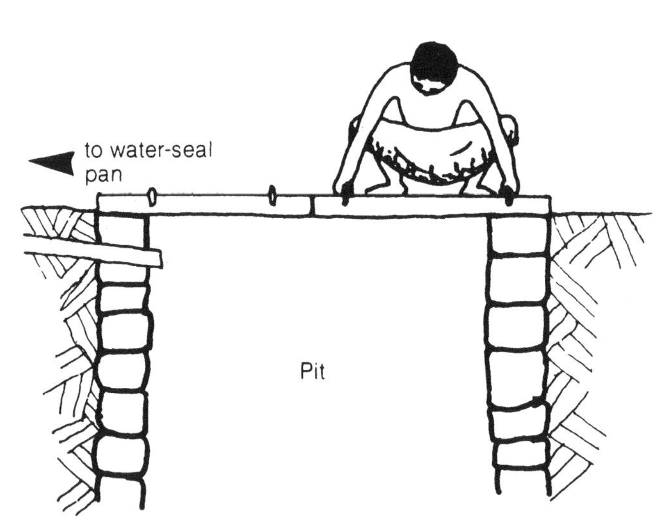

In the case of the mound latrine, first line the sides of the pit and build the lining up to 4ft above ground level.

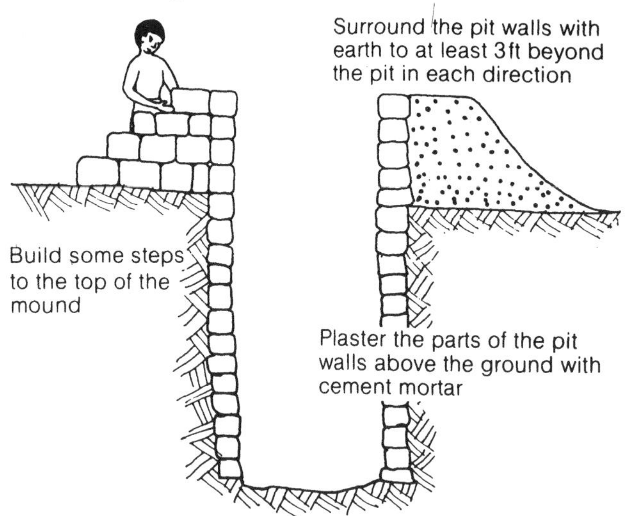

CHOOSING A WATER-SEAL LATRINE

Latrine hut

The latrine hut is the least difficult part of the latrine to make. Construct it to local preference.

Caution!

It is very important to build latrines as far as possible from wells, as they can contaminate them.

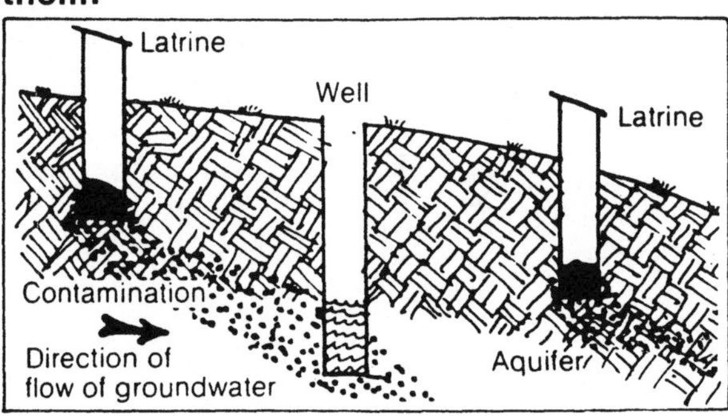

Guidelines on use and maintenance

The latrine does not have to be flushed with clean water. You can use water that has been used for washing, or water that is too salty for drinking.

To flush the bowl successfully, *hurl* the ➤ water down

Wash your hands (using soap and a brush) after using the latrine, and make sure children do as well ▼

Avoid putting any solids at all down the bowl (eg paper and leaves)

Keep the slab clean. Scrub it with a brush and use disinfectant or soap. ➤

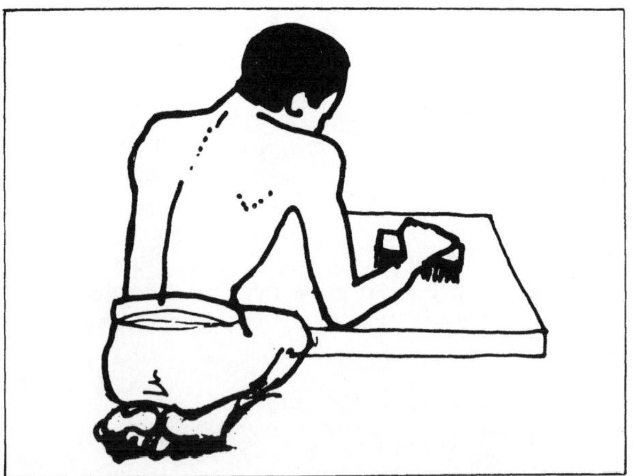

Text: Andrew Cotton, WEDC Group, Loughborough University of Technology, UK.
Illustrations: Susan Ball, WEDC Group, Loughborough University of Technology, UK.

For further information

1. Cotton, A. P. *Pit latrines for Sri Lanka*, Report for the National Housing Development Authority of Sri Lanka, 1985. Available from WEDC, Loughborough University of Technology, Loughborough, Leicestershire LE11 3TU, UK.
2. Feachem, R. and Cairncross, S. *Small excreta disposal systems*, Ross Bulletin No 8, The Ross Institute, Keppel Street, Gower Street, London WC1, UK, 1978.

7. The water cycle

The Water Cycle

Of all the reserves of water in the world, all but 0.78 per cent are tied up as either sea water or fresh water in the polar ice caps. However, this remaining amount still represents 2,360,480 cubic m of fresh water for each person in the world today. Even the rainfall alone represents 23,486 cubic m per person annually over the world's land surface. The detailed figures below give some idea of the enormous reserves of water that are available for human consumption.

		Per cent	cubic km
Total world water		100	1,386,000,000
Salt water		97.5	1,350,000,000
Fresh water		2.5	35,000,000
Comprising:	Polar ice caps	68.7	24,100,000
	Groundwater	30	10,500,000
	Comprising:		
	Lies up to 50m deep	75	
	Lies over 50m deep	25	
	Economically extractable	40	
	Lakes	0.26	176,000
	Rivers & marshes	0.03	12,000
	Soil moisture	0.05	17,000
	Atmosphere	0.04	13,000
Rainfall	On land	20	109,000
	Over sea	80	458,000
Evaporation	From land	12.5	72,000
	From sea	87.5	505,000
River run-off			45,000

Twenty per cent of water used is groundwater, and 80 per cent of water used is groundwater, and 80 per cent of water used is employed in agriculture.

The water on the earth is part of a continuous cycle (the Hydrological Cycle) with precipitation consisting of rain, snow, hail, mist and dew falling onto the earth's surface. It then either runs off over the surface into streams and rivers, or it infiltrates the ground and percolates through the ground to springs or rivers or until it reaches the ocean underflow. Water in the ocean as well as surface water on the land is turned to water vapour by the heat of the sun and so evaporates into the atmosphere. Water being used by plants and crops is also transpired into the atmosphere. This water vapour then condenses into clouds and under the right atmospheric conditions falls again to earth as rain.

The water cycle which is illustrated in this Technical Brief can also be written as an equation:

Precipitation − Evaporation − Transpiration = Run-off ± Groundwater outflow ± Change in storage

This equation can be used to determine the water resources of individual catchment areas or river basins. This is then used to ensure that only the correct amount of groundwater is taken out in order to prevent depletion of the ground-water over a number of years.

Source: World water balance and water resources of the earth, UNESCO, 1978.

Water quality

When considering how to improve a water supply it is necessary to consider the quantity, proximity and quality of the water. By enabling people to have more water, closer to home, significant health benefits can result, but in the long term it is vital to ensure that the water is also of a suitable quality.

Water easily collects all kinds of impurities which give it colour, odour, taste and turbidity (cloudiness). These impurities are either organic, derived from the decomposition of plants and animals and wastes, or inorganic such as soils, minerals and dissolved metals.

Drinking water should as far as possible be colourless, odourless and pleasant to taste. It should be free from disease-producing organisms and dissolved minerals which make the water 'hard'.

Guidelines for water quality are given below but it should be noted that whereas rainwater in rural areas will generally be of good quality if it is properly caught and stored, surface water is very likely to have been polluted and groundwater should be free from disease-causing organisms if taken from below a depth of 10m.

In unpiped water supplies, the bacteriological objective is to reduce the coliform count (statistically determined as the most probable number of a certain bacteria) to less than 10 per 100ml and to ensure the absence of faecal contamination (WHO 1984).

The chemical characteristics of rural water are not normally harmful apart from excessive fluoride and nitrate levels. However, they can cause people to reject improved water sources: for example, if iron in groundwater causes staining of clothes. It has also been noted that the hard water more typical of groundwater sources can lead to 42 per cent increase in cooking time — leading to a 44 per cent increase in the use of fuel.

mg/litre	Calcium	Chloride	Fluoride	Iron	Magnesium	Nitrate	Sodium	Sulphate	pH
Guideline values (WHO 1984)		250	1.5	0.3		10	200	400	6.5-8.5
Maximum permissible (WHO 1971)	200	600	2.0	1.0	150	100	400	400	6.5-9.2
Rainwater Example: S E Asia	2	3		0	2	1		4	5.6
Surface water Example: Sudan	7.2	10	<1	0.6	5.9	0.27	19	0.02	7.42
Groundwater Example: Granite USA	14	5.9		0	4.7		22	52	

Illustrations: Susan Ball, WEDC Group, Loughborough University of Technology, UK.
Text: Richard Franceys, WEDC Group, Loughborough University of Technology, UK.

Water demand

The Water Cycle illustrates the various sources of water and shows where water may be taken out of the cycle for domestic consumption. Detailed below is information about water usage with guidelines for design purposes.

Wells and improved sources

Source distance	Approx water usage litres per person per day
> 2,500m	5
250-2,500m	15
< 250m	15-35

The recommended design criterion is to plan for an improved water source for each population grouping of 200 to 250 people with a supply of 35 litres per person per day (45 litres per person if it is possible to design for future improvements in standard of living, but reducing to a minimum of 20 litres per day in difficult situations) within a distance of 250m of each household.

Standposts

Source distance	Approx water usage litres per person per day
< 250m	15-50 (dependent upon distance)

The recommended design values are the same as for wells. But it should be noted that approximately half to three quarters of the daily demand is likely to be consumed in the six hours between approx 6.30 am and 9.30 am and 4.30 pm and 7.30 pm.

Piped supplies

Type of supply	Approx water usage (litres per person per day)
Yard or single household tap	75
Schools: Day	25
Boarding	45
Hospitals	300 per bed
Government Offices	30
Livestock: Horses	35
Cattle	40
Pigs	15
Sheep	12.5
100 Chickens	15

To put this into perspective, the water needed for agricultural purposes for irrigation is of the order of 750 litres per kilo of grain grown and 1,500 litres per kilo of rice grown. For small scale manufacturing 100 litres of water are required per kilo of paper made, 4 litres per kilo of bread baked and 100 litres per kilo of steel made.

THE WATER CYCLE

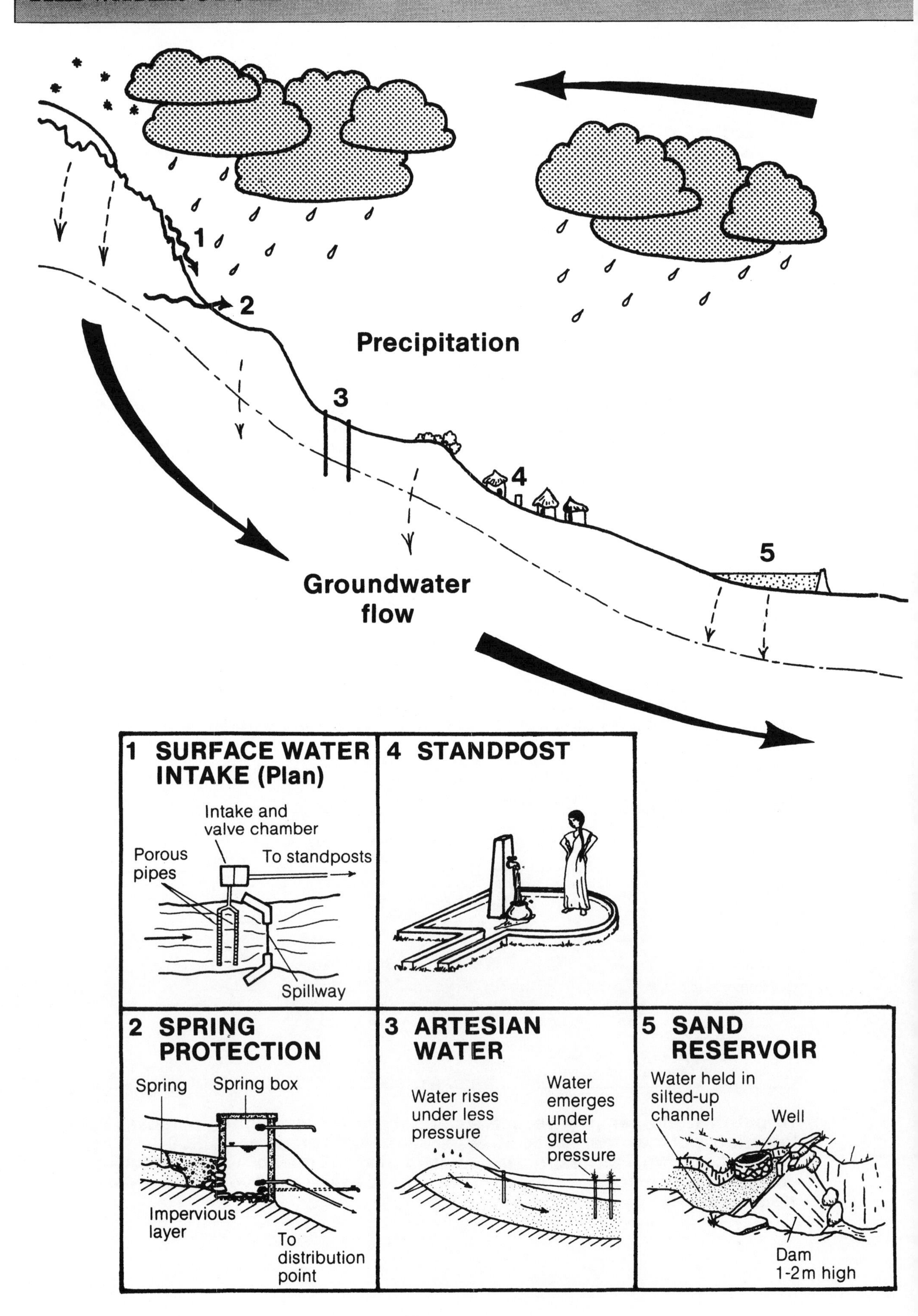

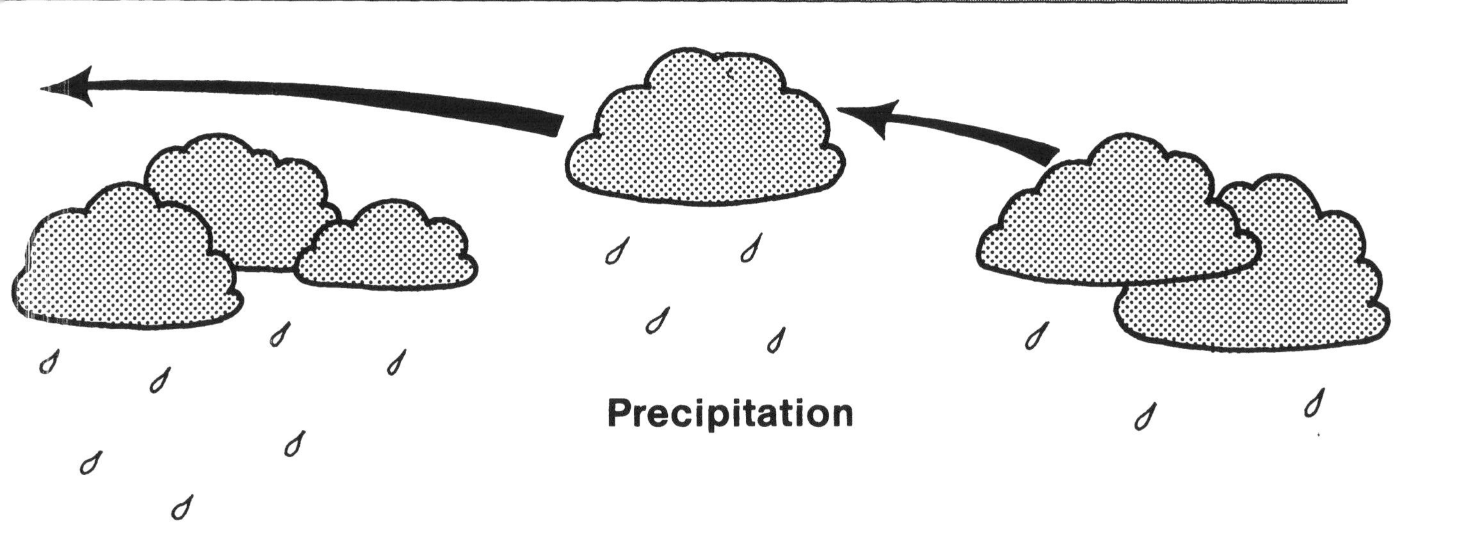
Precipitation

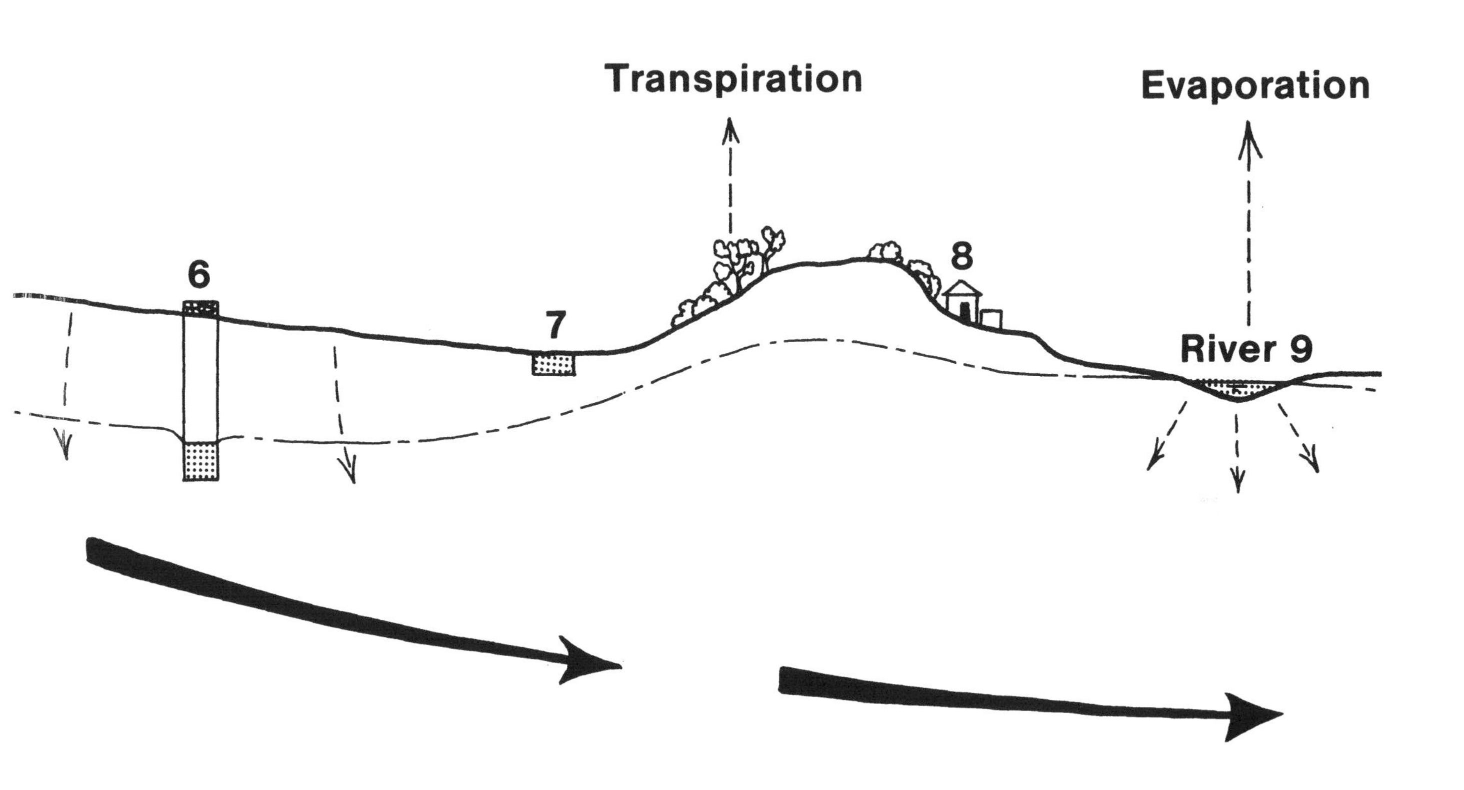
Transpiration
Evaporation
6
7
8
River 9

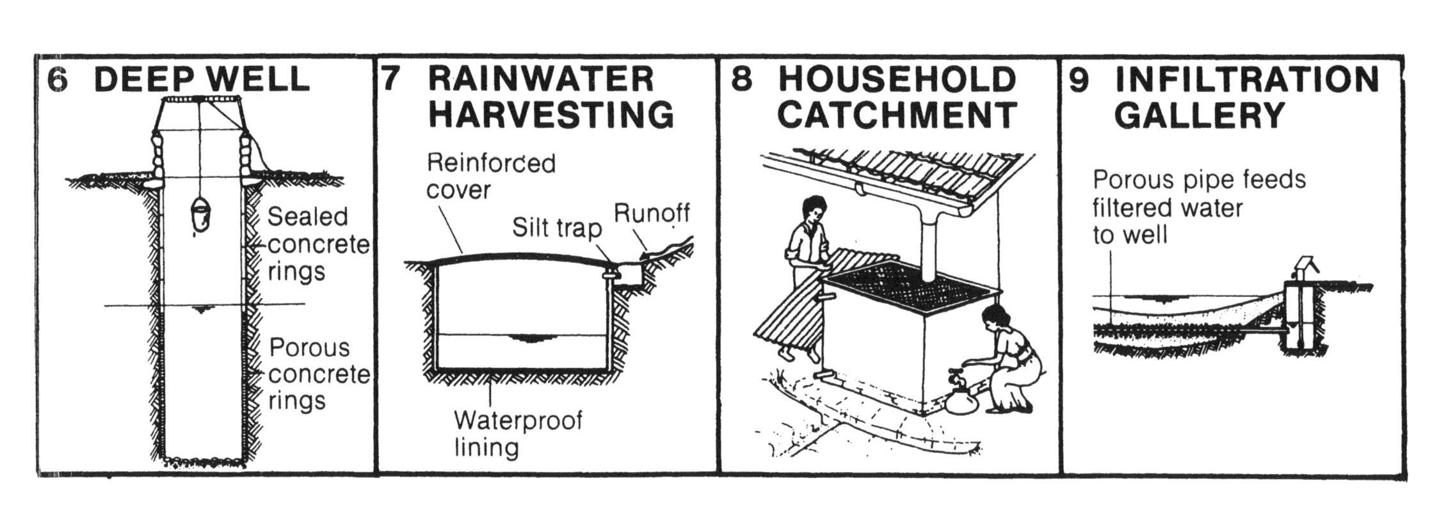
6 DEEP WELL
Sealed concrete rings
Porous concrete rings
7 RAINWATER HARVESTING
Reinforced cover
Silt trap
Runoff
Waterproof lining
8 HOUSEHOLD CATCHMENT
9 INFILTRATION GALLERY
Porous pipe feeds filtered water to well

THE WATER CYCLE

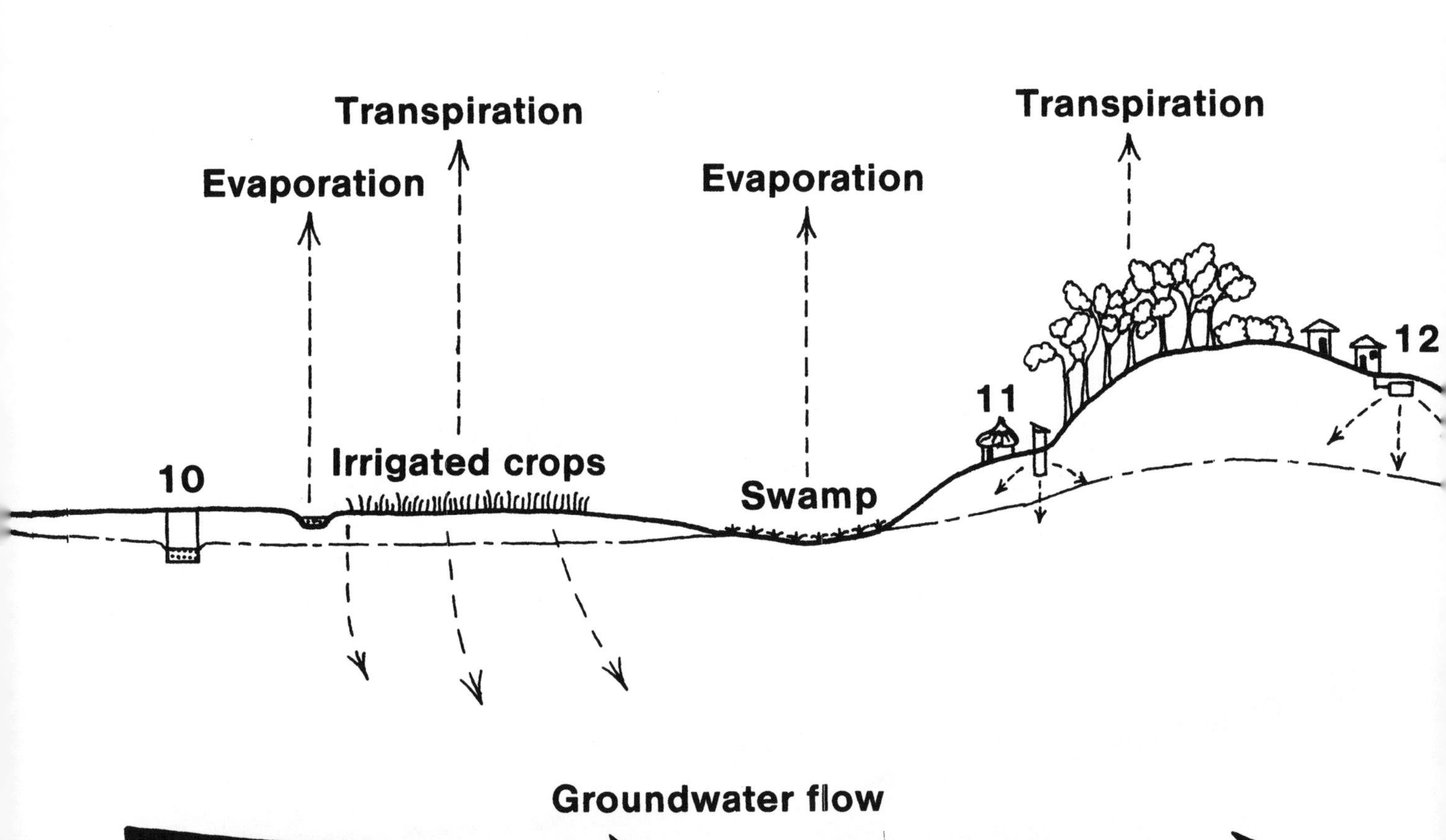

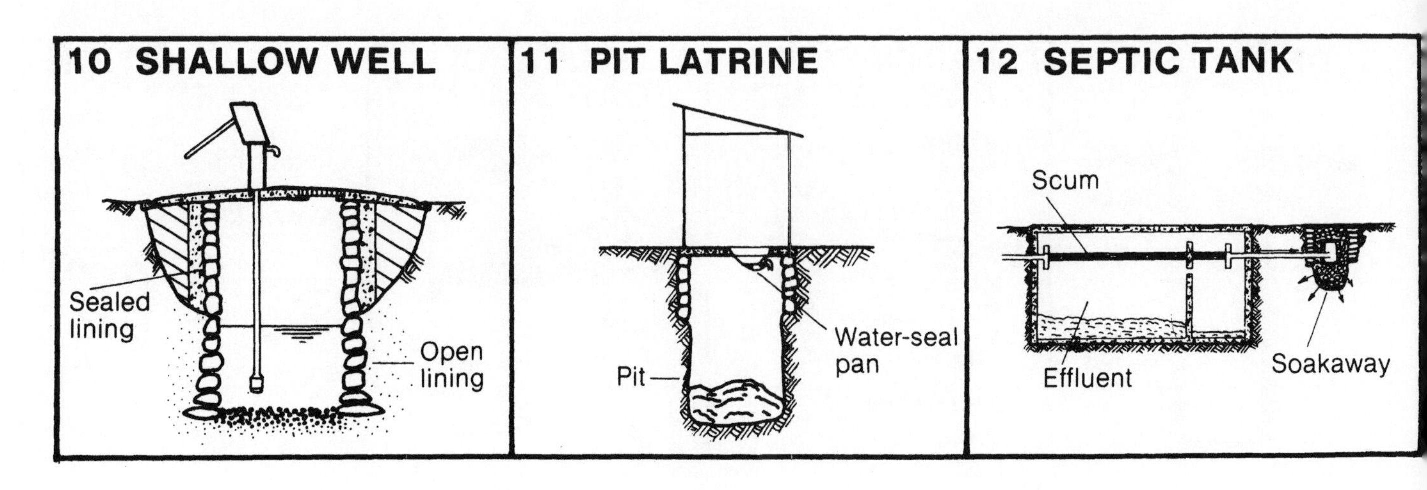

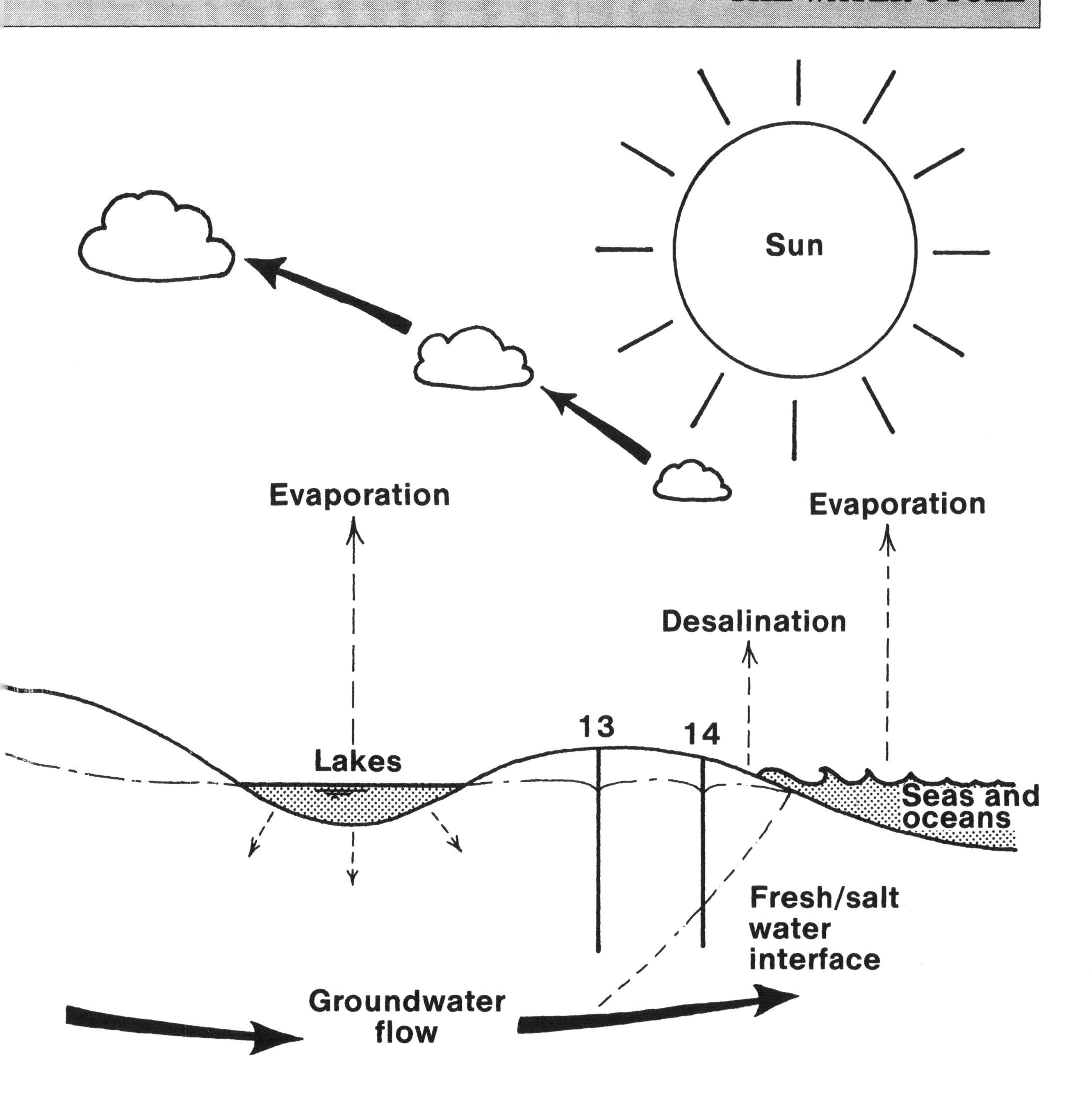
Sun
Evaporation
Evaporation
Desalination
13
14
Lakes
Seas and oceans
Fresh/salt water interface
Groundwater flow

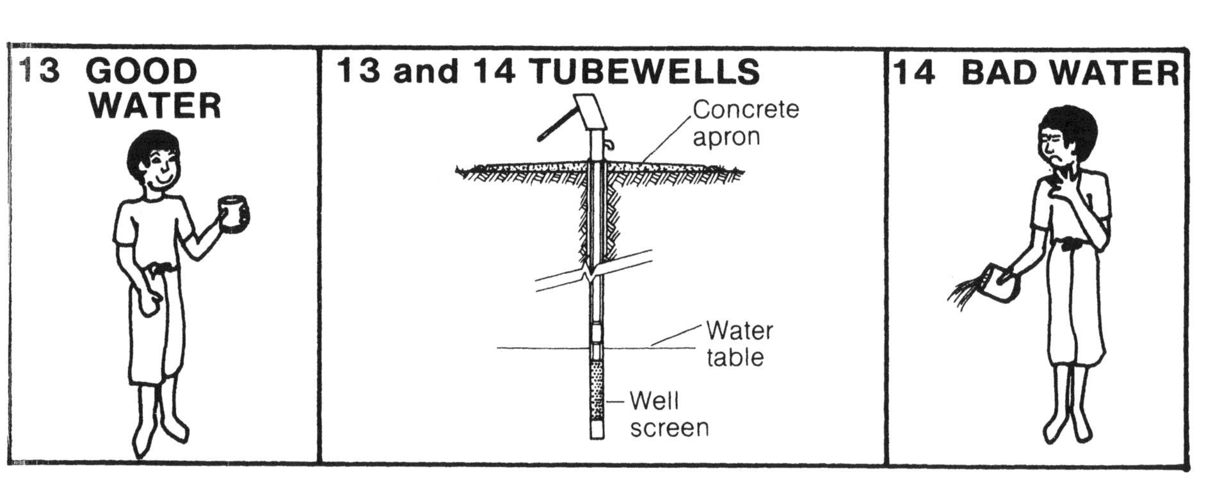
13 GOOD WATER
13 and 14 TUBEWELLS
Concrete apron
Water table
Well screen
14 BAD WATER

8. Making soap

Soap is important in preventing the spread of disease by helping people keep themselves, their clothes and their surroundings clean. In some places, soap is unavailable or expensive. This Technical Brief gives some practical guidelines on a cheap, easy way to make soap on a small scale, using ingredients which are available locally.

The principle

Making soap involves a chemical decomposition of fats and oils into their constituent parts, namely: fatty acids and glycerol. The fatty acids combine with an alkali, usually caustic soda, and the glycerol remains free. In the 'cold' process, which will be described in this Technical Brief, oil is treated with a definite amount of alkali. The aim is to complete the reaction, which generates its own heat, without any free alkali being left in the soap.

Basic recipe

To make 4 kg of soap:

- **Oil or fat – 3 litres/2.75 kg/13 cups**
- **Alkali – 370g of caustic soda crystals made up as directed on the container, or lye solution, made as described overleaf.**
- **Water – 1.2 litres/5 cups**

Choosing oils and fats

Different oils and fats bring their own specific properties to the soap, and the best mixture can only be arrived at by experimentation. Here are some guidelines, however.

***The only difference between an oil and a fat is that oils are liquid and fats are solid at normal temperatures.**

The oils and fats used in soap-making fall into three categories as shown in the table below:

An example of a suitable blend is 24 parts Category A oil, 24 parts Category B oil, 38 parts Category C fat, plus 12 parts caustic soda dissolved in 32 parts water.

***All proportions are by weight.**

Category	Composition	Type of Soap	Ratio of caustic soda:oil
A: Lauric oils eg. Coconut oil Palm kernel oil	Lauric acid is the major fatty acid	Hard soap with a fast-forming lather	1:6
B: Liquid oils eg. olive oil, corn oil, sunflower seed oil, fish oil, groundnut oil, soya bean oil, cottonseed oil	Unsaturated fatty acids	Soap lathers freely with good detergent properties, but cannot make hard soap without being mixed with other categories	1:8
C: Semi-solid fats eg. palm oil, castor oil, animal tallow	Large quantities of palmitic and stearic acids	Soap is slow to lather, but lather is more stable than that from Category A oils	1:8

Alkalis

Caustic soda is the most commonly-used alkali, but if it is too expensive or not available, caustic potash can be used. Caustic soda produces a hard soap, whereas caustic potash makes a softer soap which is more soluble in water.

To make caustic potash

A solution of caustic potash (also known as 'lye water') can be leached with water from white plant, leaf or wood ashes. The best ashes to use are those from burning hardwoods, and ashes from seaweed can also be used. Do not use ashes from burning paper, cloth or refuse.

Slowly add 7.6 litres of water to 19 litres of ashes in the apparatus shown below. After about an hour, brown lye water will start to drip from the bottom of the bucket and can be collected. When no more lye drips out, put the lye water through the ashes again to increase its strength. These quantities will make about 1.8 litres of lye.

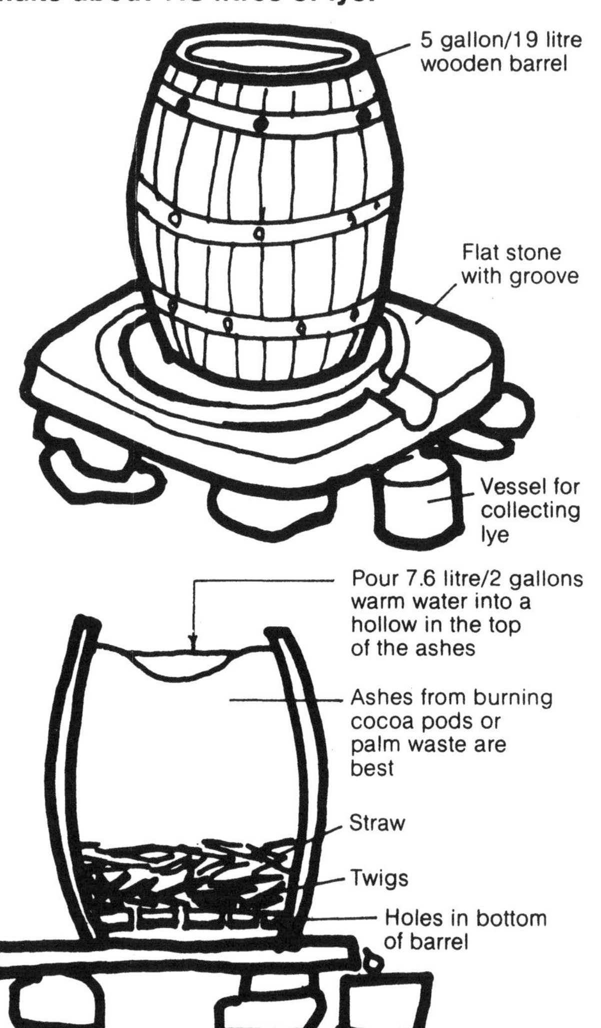

You will need 4.53 litres of lye to react with your 2.75kg of fat. It will take 48 litres of ashes to make this, according to one source.

The lye is the right strength for soap-making when it will either support a fresh egg or when it will coat, but not eat away, a chicken feather dipped into it. Concentrate the lye by boiling it, if necessary.

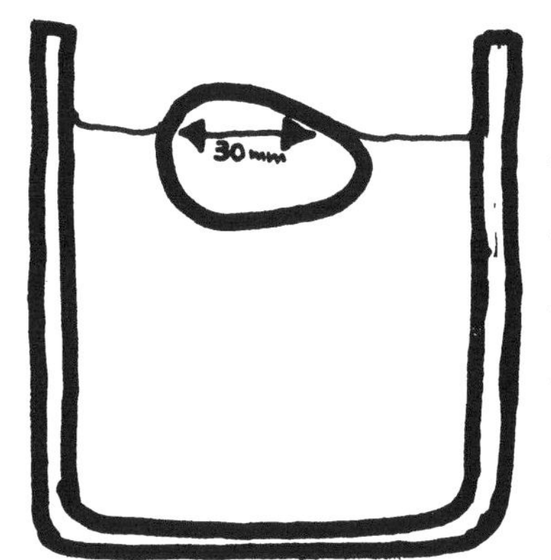

Strength of alkali

Another way of making the alkali the right strength is to make sure it is the same density as a saturated salt solution, as follows. This is equivalent to 18 per cent of caustic soda by weight (relative density 1.37).

- Dissolve a fair amount of kitchen salt in water, stir well and let it stand until the next day
- If no salt is left at the bottom, add more, stirring, until there is some left at the bottom. The solution is now saturated.
- To make a measure for the density, take a small stick of solid wood, and weight the end (by tying on a pebble or a small piece of iron). Put the weighted stick into the salt solution. Adjust the weight so that it floats with a small part of the stick protruding from the salt solution. Mark the stick where it touches the surface

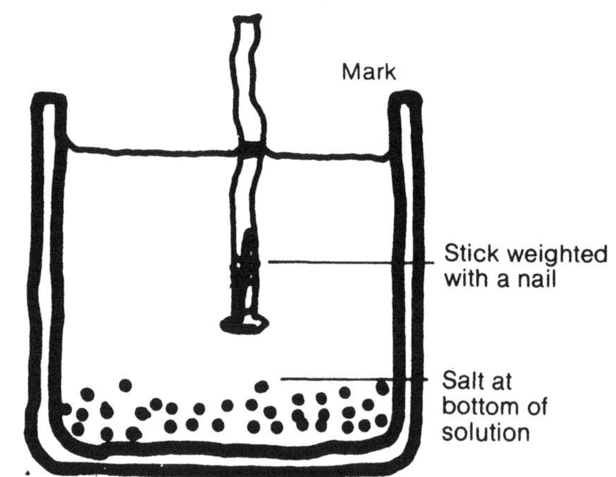

If you then put the stick in the alkali, it will float with the mark submerged if the lye is too weak. If it is too strong, the mark will be above the surface of the liquid. Adjust the strength by boiling to concentrate it or adding water, stirring well, until the mark is exactly at the surface.

MAKING SOAP

Water

Water needs to be 'soft' to make good soap, so rainwater is a good source. 'Hard' water contains dissolved mineral salts which prevent soap lathering and hinder cleansing.

To make water soft, add 15ml (1 teaspoonful) of lye to each 3.8 litres (1 gallon). Stir and leave to stand for several days, to allow the sediment which has been precipitated to sink to the bottom. Then pour off the softened water.

Equipment

To make soap, you will need:

- Two large bowls or buckets. Soap-making equipment should never be made from aluminium, as the alkali will corrode it

- Measuring cups
- Wooden or enamel spoons or smooth sticks for stirring
- Moulds: water-tight containers which can be made from wood, plastic, cardboard or waxed paper

Gourds or coconut shells also make good moulds

- Cloth or waxed paper to line the moulds, so that the soap can be removed easily

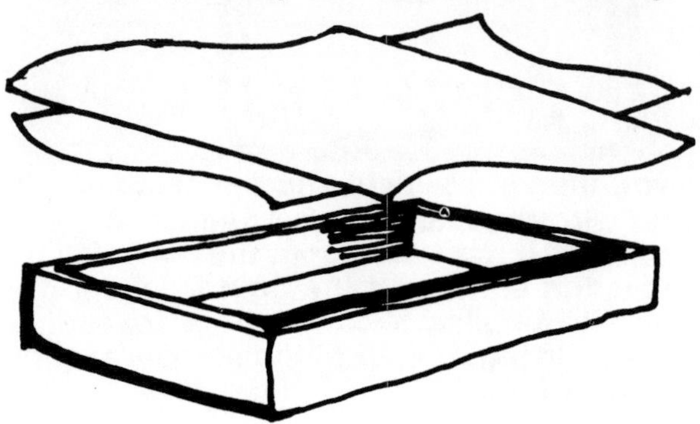

Make the lining of two strips: one longer than the mould, and the other wider

> WARNING: Caustic soda will burn skin and eyes, so try and wear protective gloves while making soap. If you get burnt, wash the skin immediately with cold water and then put citrus juice or vinegar on it to neutralize the alkali.

Method

- Add alkali to water, never the other way round. The alkali solution should be at body temperature (37°C). Never put your finger in the solution to test the temperature, or it will burn you, but feel the outside of the container

- Melt any solid fat in the oil/fat mixture
- Pour the alkali slowly into the oil/fat mixture, stirring it continuously in one direction only. The mixture needs to be stirred for at least half an hour after all the alkali has been added. The mixture should become thicker, and lines of white particles should follow the spoon as you stir
- Pour the mixture into lined moulds and leave it to set undisturbed for two days in a dry place. If it has obviously not set after two days or grease is visible on the top, leave it a little longer
- When the soap has set, remove it from the moulds and cut into bars

Cut the soap with a knife or wire

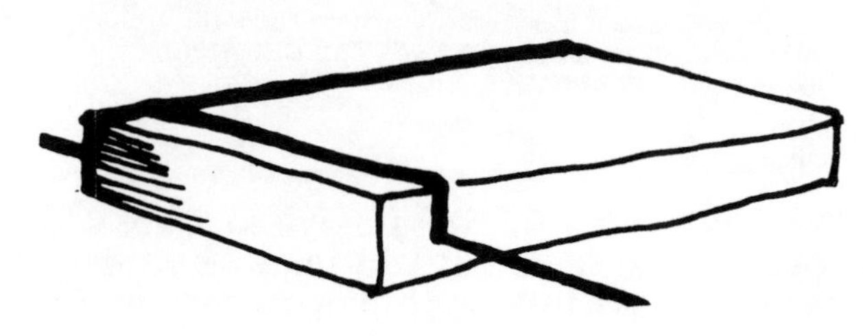

- Stack the bars on trays and leave them for four to six weeks to allow the chemical reaction to finish completely

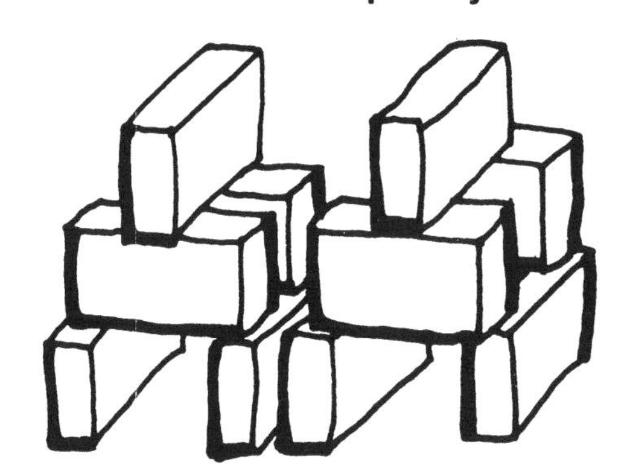

- When the soap is finished, it will shave from the bar in curls. Cover the bars of soap to prevent further loss of moisture

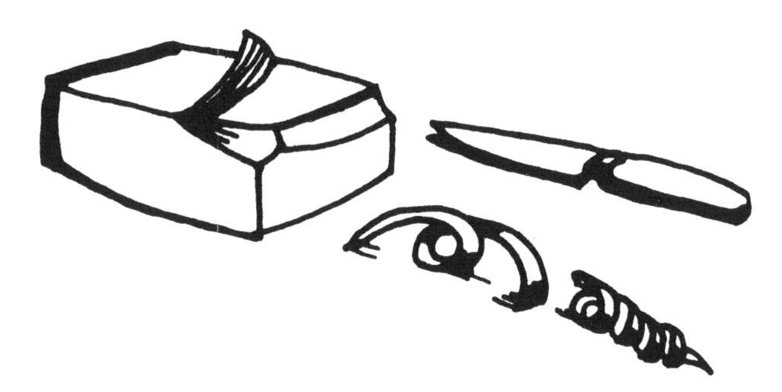

Perfume

Perfume can be added at the same stage as the alkali. As well as giving the soap an attractive smell, it can act as a preservative. Perfumes must be resistant to alkali, however. For 4 kg of soap, one of the following could be used:

- 4 teaspoons oil of sassafras
- 2 teaspoons of oil of wintergreen or citronella or lavendar
- 1 teaspoon of oil of cloves or lemon

Problems?

If the soap you made was not successful, it may have been because:

- You used the wrong materials
- The fat or oil was rancid or salty
- The alkali was too hot or cold
- The mixture was stirred too fast or not for long enough

Using dirty or rancid fat

Dirty or rancid fat must be cleaned before it can be used for making soap. This can be done by melting it and straining it through a finely woven cloth or by boiling it up with water, leaving it to cool and separating it when set.

For more information

1. Donkor, Peter. *Small-scale soap-making: a handbook*, Intermediate Technology Publications.
2. Bertram, S. P. *The preparation of soap*, TOOL, Entrepotdok 68-69a, 1018 AD Amsterdam, The Netherlands. 1976.
3. Tropical Development and Research Institute. *Soap manufacture by the cold process*, TDRI, 56-62 Gray's Inn Road, London WC1X 8LU, UK.
4. German Adult Education Association. *Make your own soap: an aid to extension and village workers in Ghana*, African Bureau of the German Adult Education Institute, PO Box 9298, 36 Patrice Lumumba Road, Accra, Ghana.
5. *VITA Village Technology Handbook*, 1815 North Lynn Street, Rosslyn, Virginia 22209, USA.

Compiled by Kathy Attawell and Katherine Miles, *Dialogue on Diarrhoea*, Appropriate Health Resources and Technologies Action Group, 85 Marylebone High Street, London W1M 3DE, UK.

Illustrations by Frances Stuart

9. Dry latrines

Where water is only available in small amounts, a simple dry pit latrine is a very good sanitation option. A well constructed latrine has the following features:

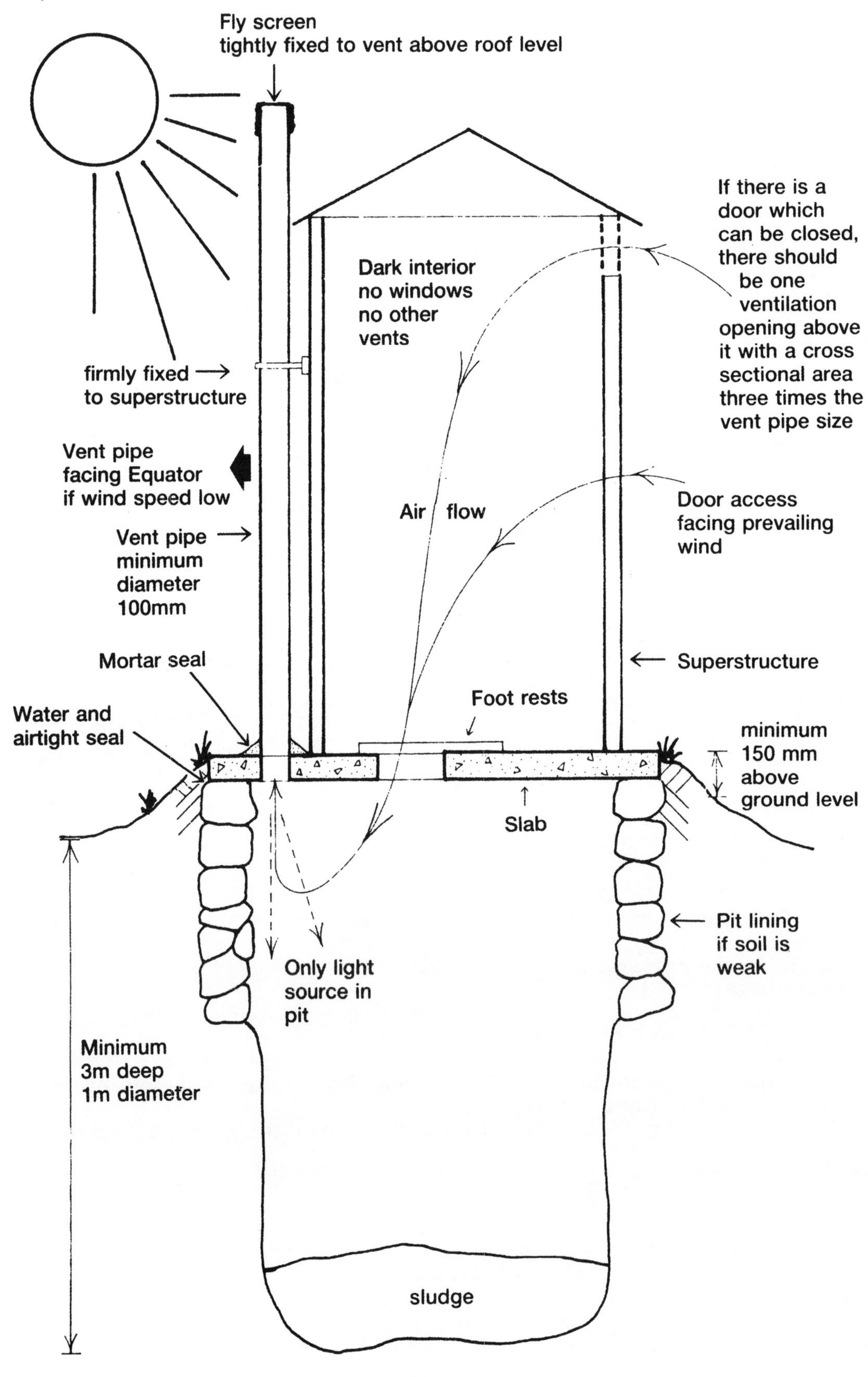

Pit size

This will depend on how many people are using the latrine, what they use for anal cleansing and the height of the groundwater table.

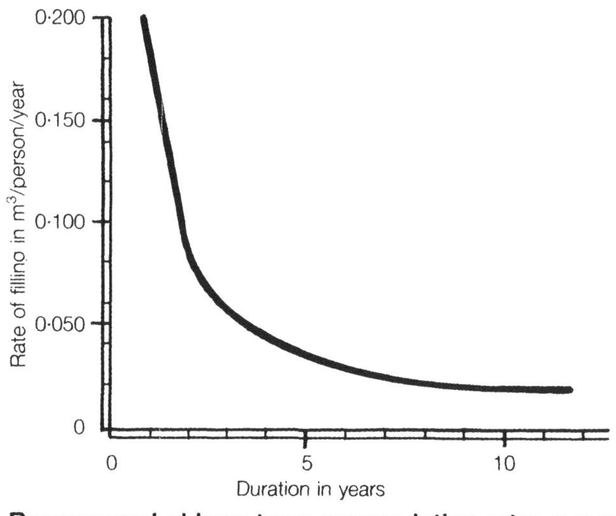

A pit is considered full when the sludge inside has risen to within 0·5m of the slab. Because the sludge in the pit digests naturally over time, the rate of filling declines the longer the pit is used.

Recommended long-term accumulation rates per person per year:
a) below water level with biodegradable anal cleansing materials – 0·020m^3 to 0·040m^3
b) dry conditions with biodegradable anal cleansing materials – 0·040m^3 to 0·060m^3

Pit shapes

Round pits have stronger walls

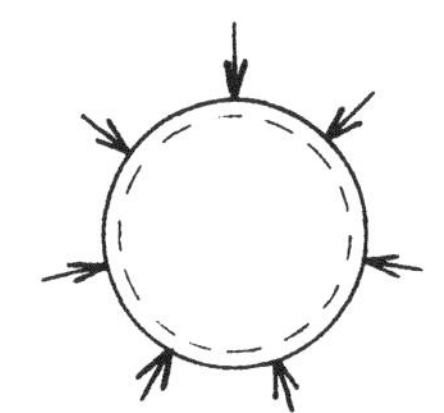

Rectangular pits are easier to dig but there is more danger of collapse.

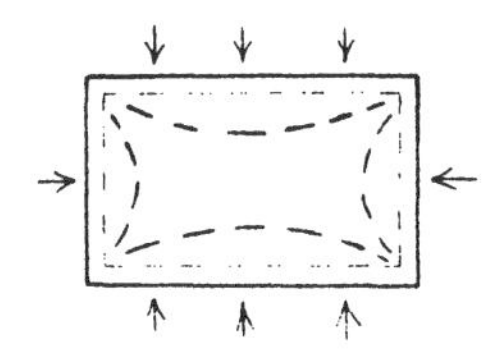

Deep pits will last longer than shallow pits of similar volume.

Pit linings

Many different materials can be used depending upon the ground conditions and local availability.

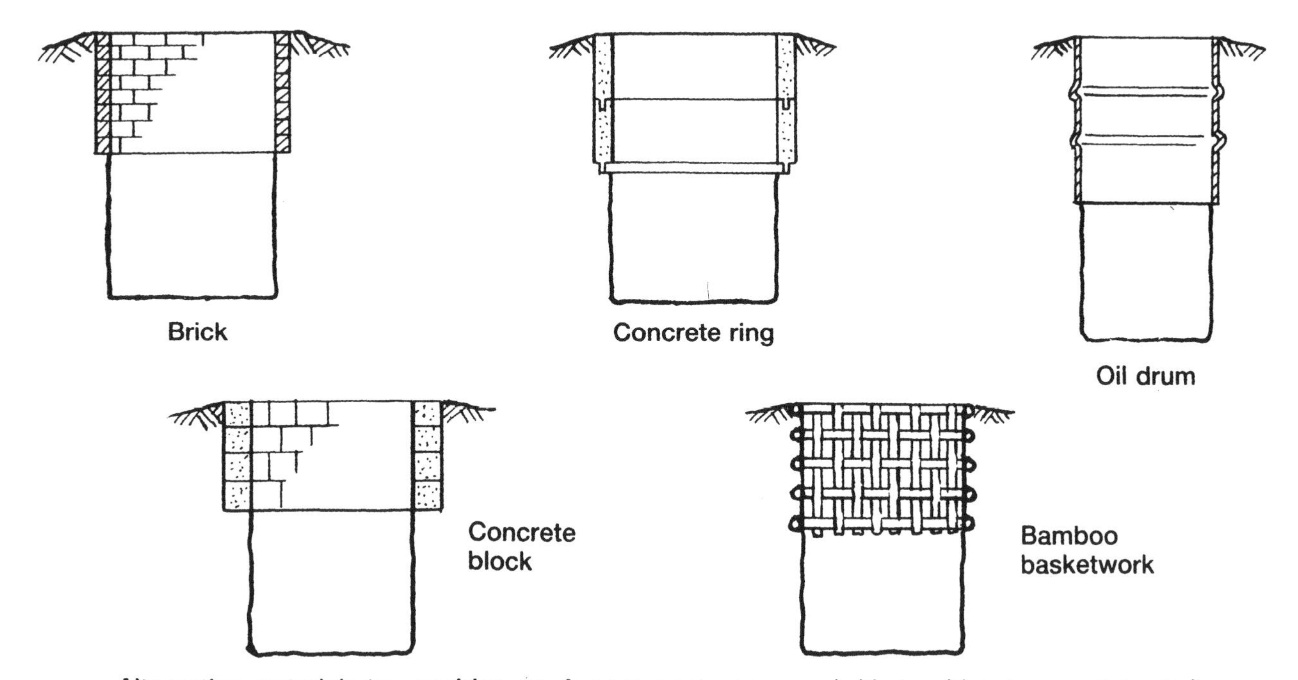

Alternative materials to consider are: ferrocement, masonry (with or without cement mortar), burnt clay ring, stabilised mud block etc.

Examples of latrine slabs

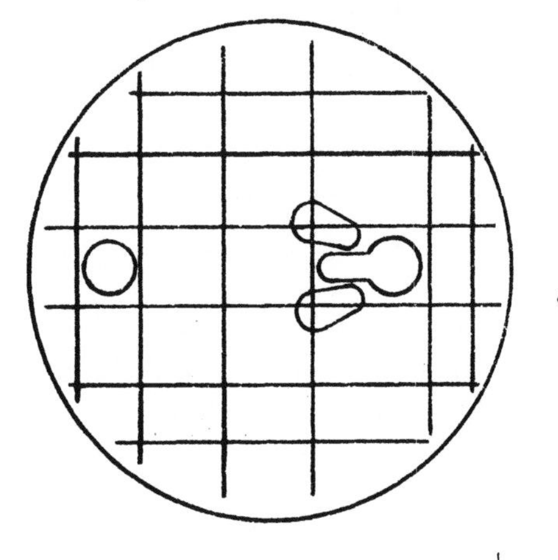

Reinforced Concrete

Concrete mix:
Cement – 24 litres (or 2/3 of 50kg bag)
Sand – 48 litres
Gravel – 96 litres (6-20mm size)
Water – 20 litres
14m of 6mm Rebar
This slab weighs about 275kg
and can be rolled into position.

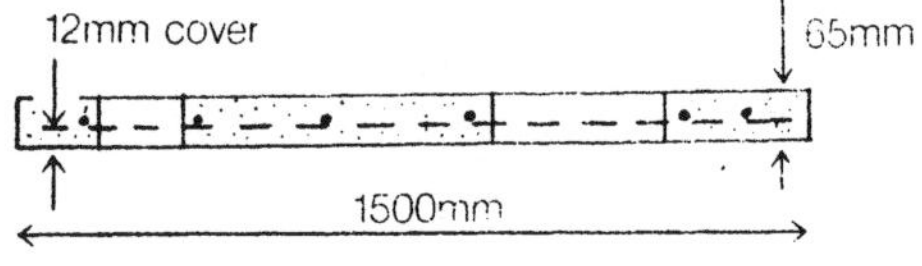

A rectangular slab can be pre-cast in two pieces to reduce the weight approximately 180kg each.

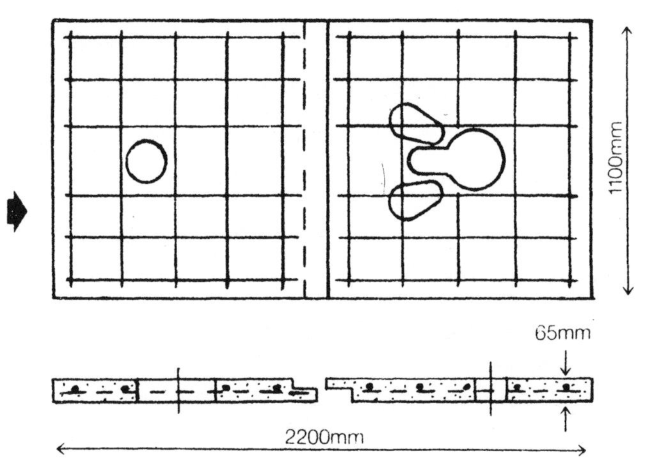

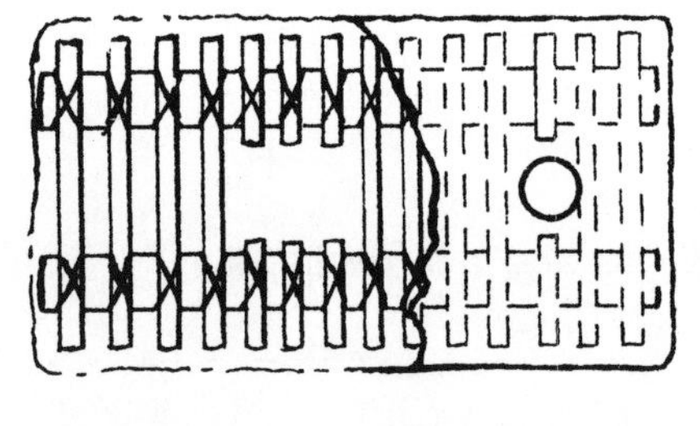

Local Materials

Wood or bamboo lashed together and covered with a layer of mud makes a strong slab where other materials are difficult to acquire.

Unreinforced Concrete

A simple concrete domed slab without reinforcement is another alternative. No vent is needed but a close-fitting plug must be provided to control flies and odour.

The shape of the slab can be made by mounding up earth within a circular former made from a strip of steel.

This type of slab requires 2/3 of a 50kg bag of cement, and weighs about 275kg.

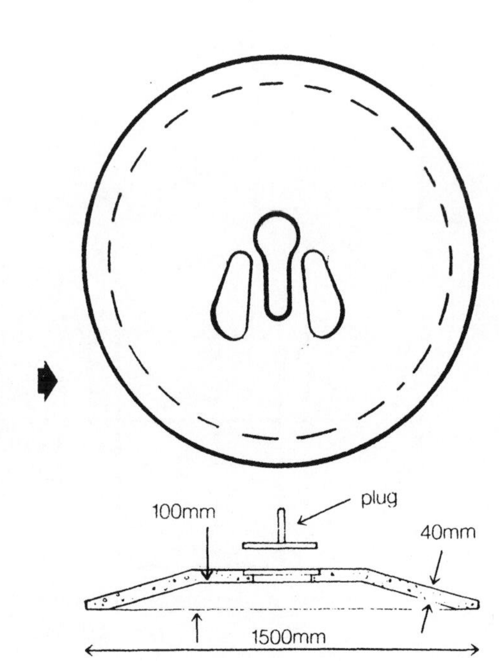

Superstructures

A superstructure can be built from locally available material eg:

Mud and thatch

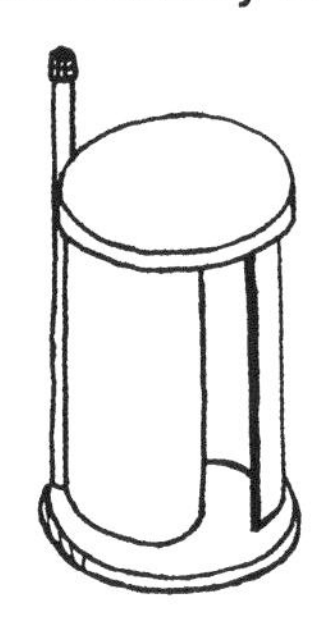

Ferro-cement

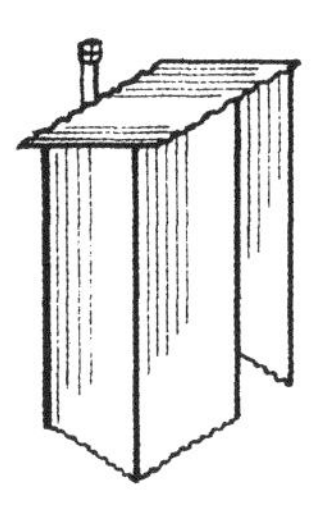

Galvanised corrugated iron

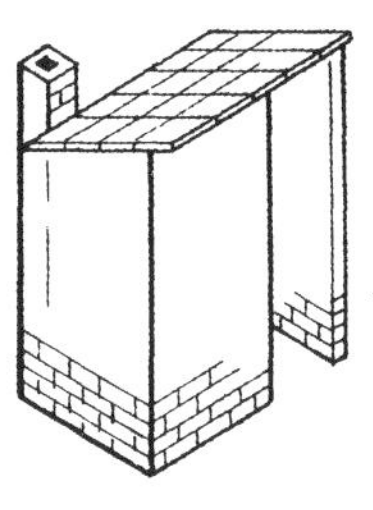

Tiles and brick

It may be of various shapes and orientation with a round or rectangular pit and slab:

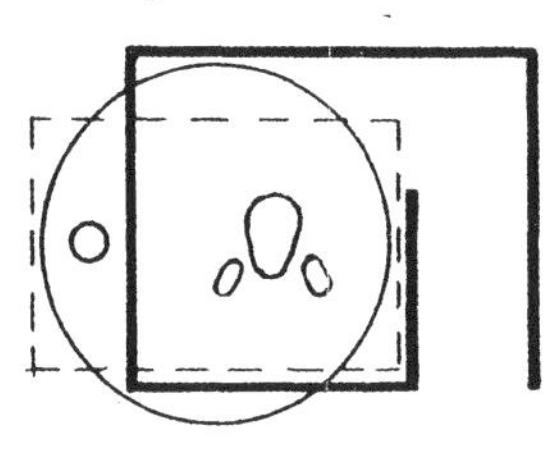
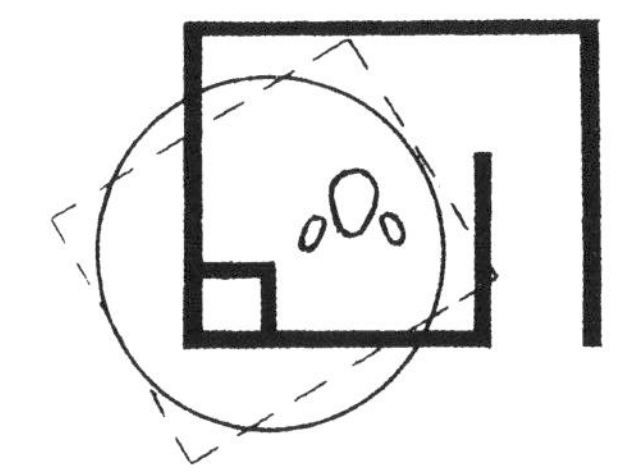
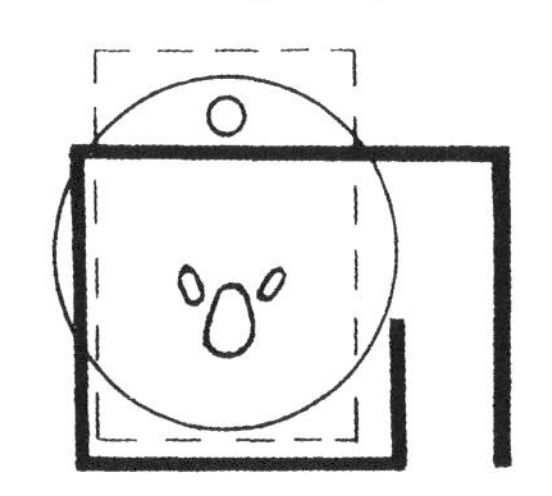
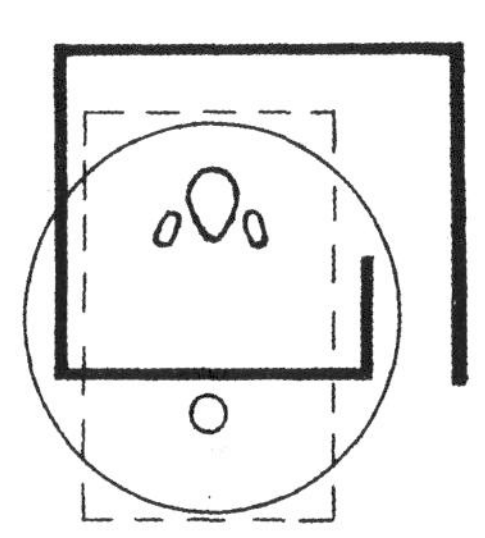

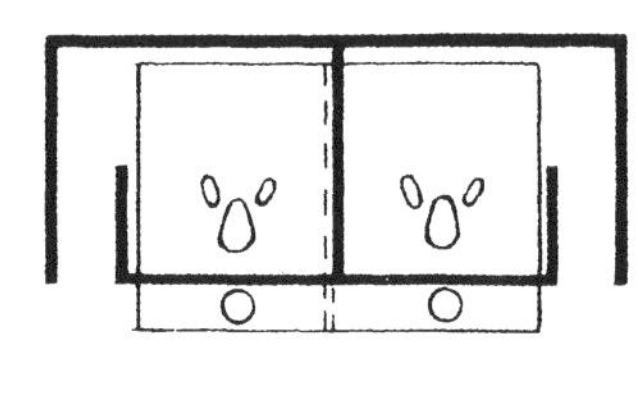
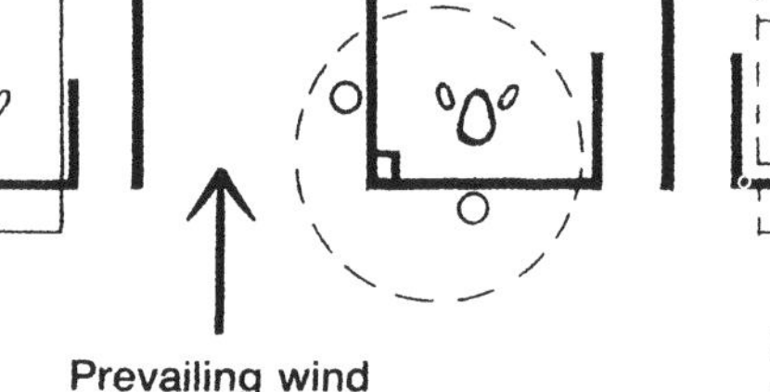

Prevailing wind direction

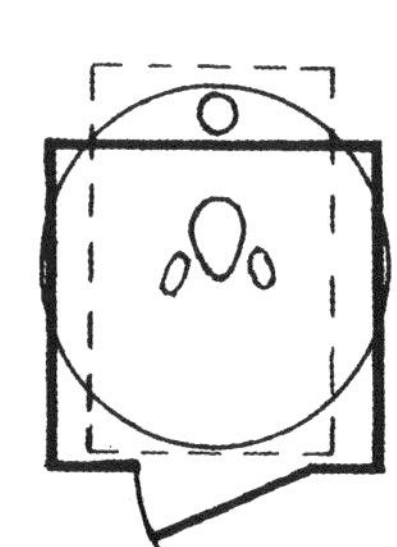

Many latrines are built like this, but the spiral design controls flies and odour more effectively.

Vents

Brick chimney – constructed as part of the superstructure, either in one corner or centre of external wall. 180mm to 230mm internal diameter.

Cement plastered hessian over chicken wire frame. 200mm to 250mm internal diameter.

Cement plastered split bamboo or reeds. 200mm to 250mm internal diameter.

UV Resistant Plastic or Asbestos Cement Pipe. 100mm to 150mm diameter.

(In each case the larger vent size should be used if mean wind speed is below 3m/sec)

Fly screens

A screen is required at the top of the vent to stop flies entering and escaping from the pit. Flies which are attracted by the light will die as they try to pass the screen and will then fall back into the pit. To minimise losses in the air flow the openings in the screen should not be smaller than 1.2mm by 1.5mm.

Ordinary mosquito wire will corrode very quickly because of gases in the pipe. PVC coated glass fibre, or for extra life, stainless steel should be used.

(For more information on vents and screens read TAG Technical Note No 6, Ryan and Mara, 1981)

For further information:

Wagner, E.G. and Lanoix, J.N. *Excreta disposal for rural areas and small communities,* WHO Monograph 39, Geneva 1958. (Under revision).

Richard Franceys, WEDC, Loughborough University of Technology, UK.
Susan Ball, WEDC, Loughborough University of Technology, UK.

10. Waste stabilisation ponds

Waste Stabilisation Ponds (WSPs) are large, shallow, man-made lakes in which bacteria help to purify raw sewage.

A possible system of waste stabilisation ponds using several types of pond might look like this:

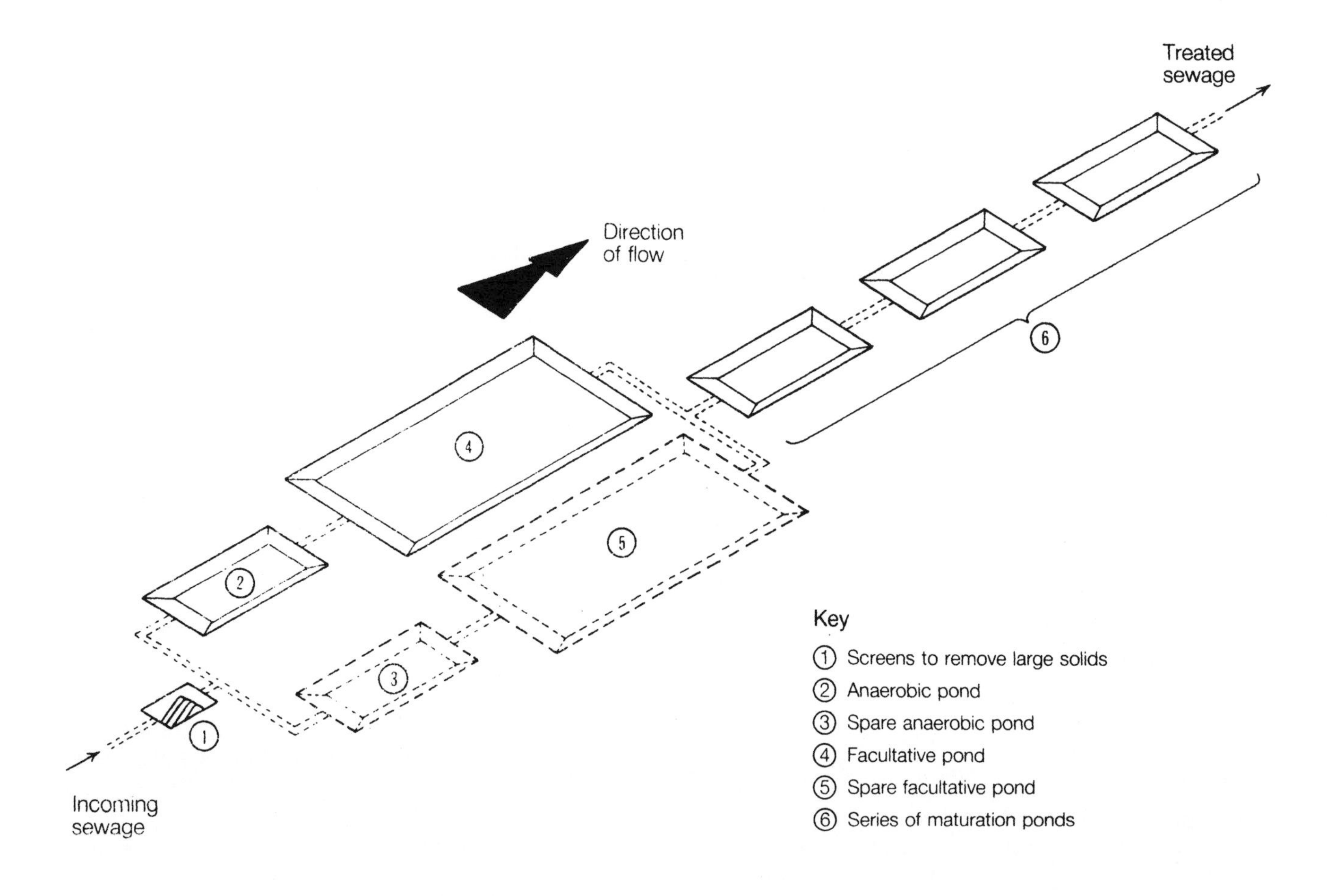

So that any pond may be isolated for maintenance work or to allow sludge to be removed, at least one extra pond of each type should be provided.

This system is suitable for treating sewage (household wastewater and excreta) from quite large communities. Possible applications are:

1. Treatment of sewage collected by a network of sewers,
2. Treatment of sewage collected in small-bore sewers,
3. Treatment of nightsoil collected from a community.

WSPs take up large areas of land and so are suitable only where land is easily available.

Suggested design methods

1. Anaerobic ponds

Volumetric BOD loading rate (grams O_2/m^3. day)	$= \lambda_v$
Influent BOD strength (mg. O_2/litre)	$= L_i$
Influent flow rate (m³/day)	$= Q$
Pond volume (m³)	$= V$
Mean temperature of the coldest month (°C)	$= T$

$$\lambda_v = \frac{L_i Q}{V}$$

Find the appropriate value of λ_v from the following and calculate V.

$T < 10°C$, $\lambda_v = 100$ grams O_2/m^3. day
$20°C > T > 10°C$, $\lambda_v = (20T—100)$ grams O_2/m^3.day
$T > 20°C$, $\lambda_v = 300$ grams O_2/m^3.day

2. Facultative ponds

Surface BOD loading (kg O_2/ha. day)	$= \lambda_s$
Mid-depth plan area of pond (ha)	$= A_f$
Influent BOD strength (mg. O_2/litre)	$= L_i$
Influent flow rate (m³/day)	$= Q$
Mean depth of pond (m)	$= D_f$
Mean retention time (days)	$= t_f$
Mean temperature of the coldest month (°C)	$= T$

Calculate λ_s from

$$\lambda_s = 350\,(1.107 - 0.002T)^{(T-25)}$$

Calculate $A_f = \frac{10 L_i Q}{\lambda_s}$ or $t_f = \frac{10 L_i D_f}{\lambda_s}$

3. Maturation ponds

No of faecal coliforms in effluent (No/100ml)	$= N_e$
No of faecal coliforms in influent (No/100ml)	$= N_i$
First order faecal coliform removal constant (days⁻¹)	$= K_b$
Mean retention time in a pond (days)	$= t$
Mean temperature of coldest month (°C)	$= T$

$$N_e = \frac{N_i}{\text{(Product of } (1 + K_b t) \text{ for } all \text{ ponds in series)}}$$

(Note: Anaerobic, facultative and maturation ponds are all considered in calculating the product of $(1 + K_b t)$.

$$K_b = 2.6\,(1.9)^{(T-20)}$$

Pond type	Typical Depth	Typical Retention Time
Anaerobic	2-5 metres	3-5 days
Facultative	1-2 metres	20-40 days
Maturation	1-2 metres	4-6 days for each of three or more ponds

The purification process

In very simple terms, incoming sewage which has usually been passed through metal screens to remove large solids, enters a system of ponds. Some of the wastes float to the surface as scum, while other wastes sink to the bottom as sludge. The first pond in a series would look like this, in section:

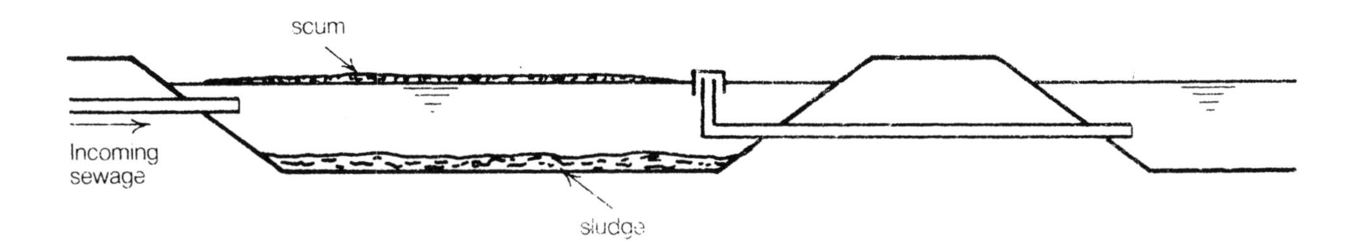

Over a period of time, bacteria living in the ponds feed on the wastes, partially treating them.

Sunlight is needed to encourage the growth of algae which are essential to the purification process in facultative ponds. Warm temperatures accelerate the treatment of wastes, and wind is important to ensure good mixing of the pond contents. WSPs work well in hot climates.

Pond form and layout

Ponds are often rectangular in plan, with depths varying from 1 to 6 metres. There are three types of pond which may be used:

1. Anaerobic ponds – used to pre-treat strong wastewaters (not always needed).
2. Facultative ponds – used to break down the organic matter in the sewage.
3. Maturation ponds – used to destroy faecal pathogens.

A typical System might be:

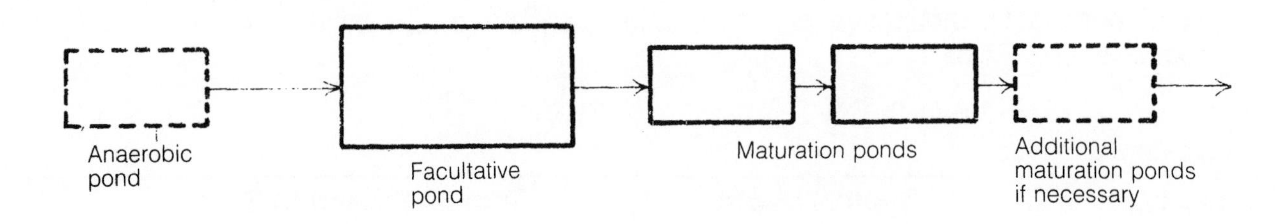

Anaerobic ponds perform the same function as septic tanks, so are not necessary if the sewage comes from septic tanks along small-bore sewers.

Design features

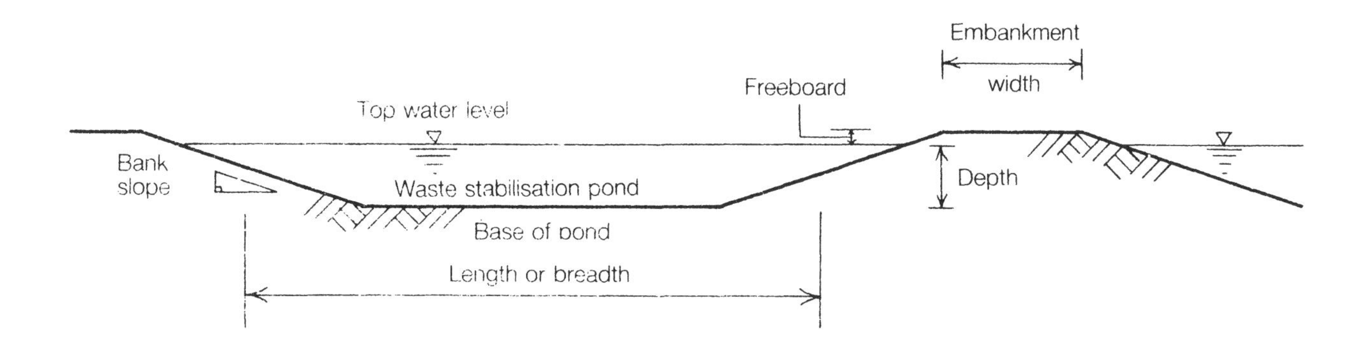

Depth: Typical ranges for pond depths are:

Anaerobic ponds	2.0 to 5.0 metres
Facultative ponds	1.0 to 2.0 metres
Maturation ponds	1.0 to 2.0 metres

Length or breadth: It should be assumed that the length and breadth of a pond are measured at half depth.

Freeboard: A clear height of 0.5 metres should be provided between the top water level and the top of the embankment.

Bank slope: Embankment slopes should be at about 1:3.

Embankment width: Maintenance vehicles should be able to have access between adjacent ponds.

Base of pond: Pond bases should be impermeable, lined with clay, plastic, rubber or concrete, to prevent leakage and groundwater contamination.

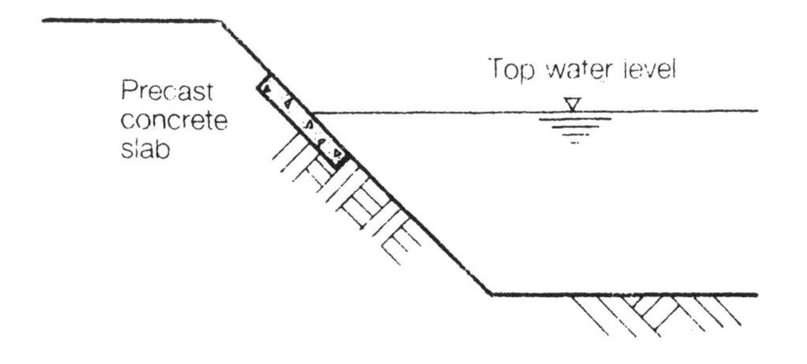

Pre-cast concrete slabs should be laid at top water level around each pond to discourage growth of weeds and to prevent bank erosion caused by wave action.

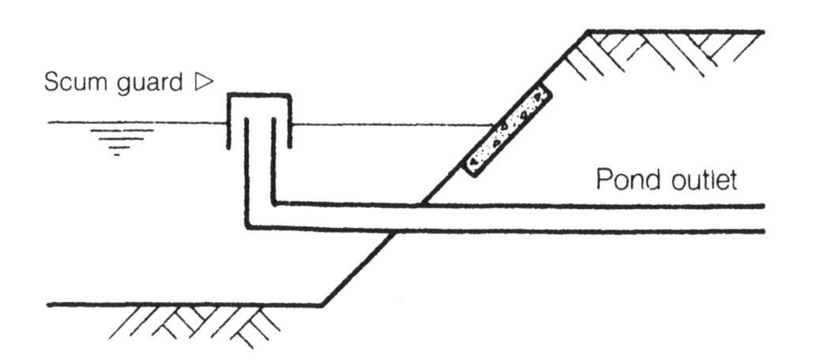

Scum guards should be provided around outlets from all ponds to prevent floating material from entering and possibly blocking the pipes.

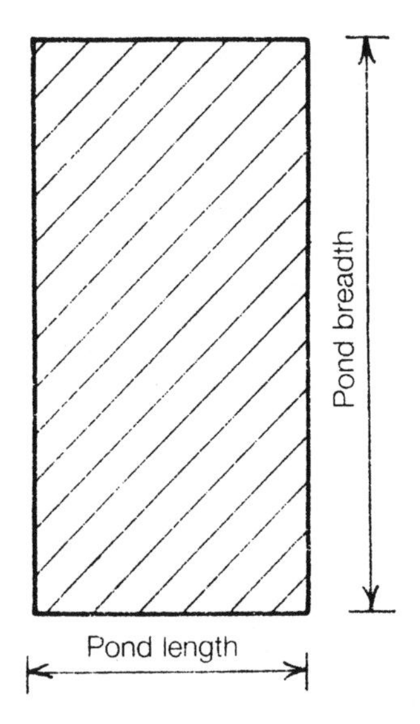

In order to ensure that the pond contents are well mixed by winds the ratio of breadth to length should be in the range 1:2 or 1:3.

The inlet to a series of ponds should be constructed in concrete to prevent bank erosion, and should be designed so that sewage enters below top water level.

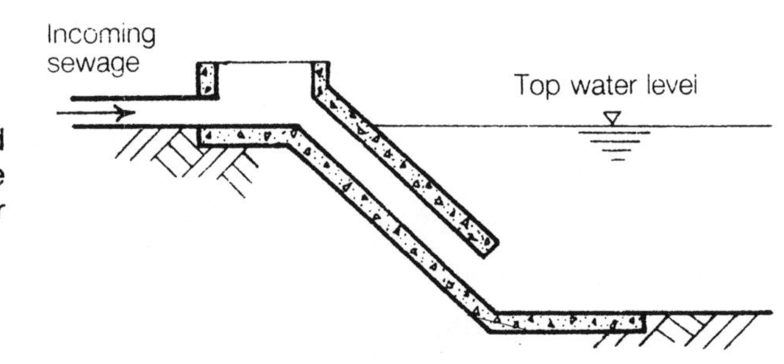

Maintenance

1. Any scum that collects on the pond surfaces should be removed and either buried or burned.
2. If screens or grit traps are used to collect the easily separated solid materials at the inlet to the ponds, the materials collected should be buried.
3. Grass around the ponds must be cut regularly.
4. Anaerobic and facultative ponds will require desludging every few years as necessary. (Anaerobic ponds every 3 to 5 years; facultative ponds every 10 to 15 years).
5. Some bird scaring may be necessary in order to reduce the likelihood of bird droppings polluting the partially treated sewage and also to reduce cross-pollution between ponds.

For further information:

Horan N.J. *Biological Wastewater Treatment Systems: Theory and operation*, John Wiley & Sons, Chichester, England, 1990.

Lumbers J. and Andoh B. *Waste stabilisation pond design*, Waterlines Vol. 3 No. 4, 1985.

Mara D.D. *Sewage treatment in hot climates*, English Language Book Society/John Wiley & Sons, Chichester, England, 1977.

Mara D.D. *Sewage treatment in hot climates*, Overseas Building Note No. 174, Building Research Station, Watford, England, 1977.

Michael Smith, WEDC, Loughborough University of Technology, UK.
Susan Ball, WEDC, Loughborough University of Technology, UK.

11. Rainwater harvesting

Rainwater harvesting is a method of collecting and using precipitation from a small catchment area.

Stored rainwater can be a valuable supplement to other, possibly inadequate, domestic water sources, and also for irrigation.

Its use is particularly appropriate in parts of the world where heavy, intense storms are followed by prolonged periods of little or no rainfall.

An adequate household supply system needs a CATCHMENT AREA, a means of COLLECTION and good STORAGE facilities.

RAINWATER HARVESTING

Catchment

The catchment area may be a roof or an area of ground.

Roof catchment:
Corrugated iron roofs are often used. They are cheap, durable and easily maintained. Costs may also be kept down by the use of local materials such as clay tiles, sisal-cement tiles or thatch. However, water collected from thatch is discoloured and usually contaminated. An improvement is to use plastic sheet over the thatch. Tiled or metal roofs give the cleanest water.
Collection of rainwater from a roof is the cheapest form of rainwater harvesting since the additional costs are then limited to the collection and storage elements.

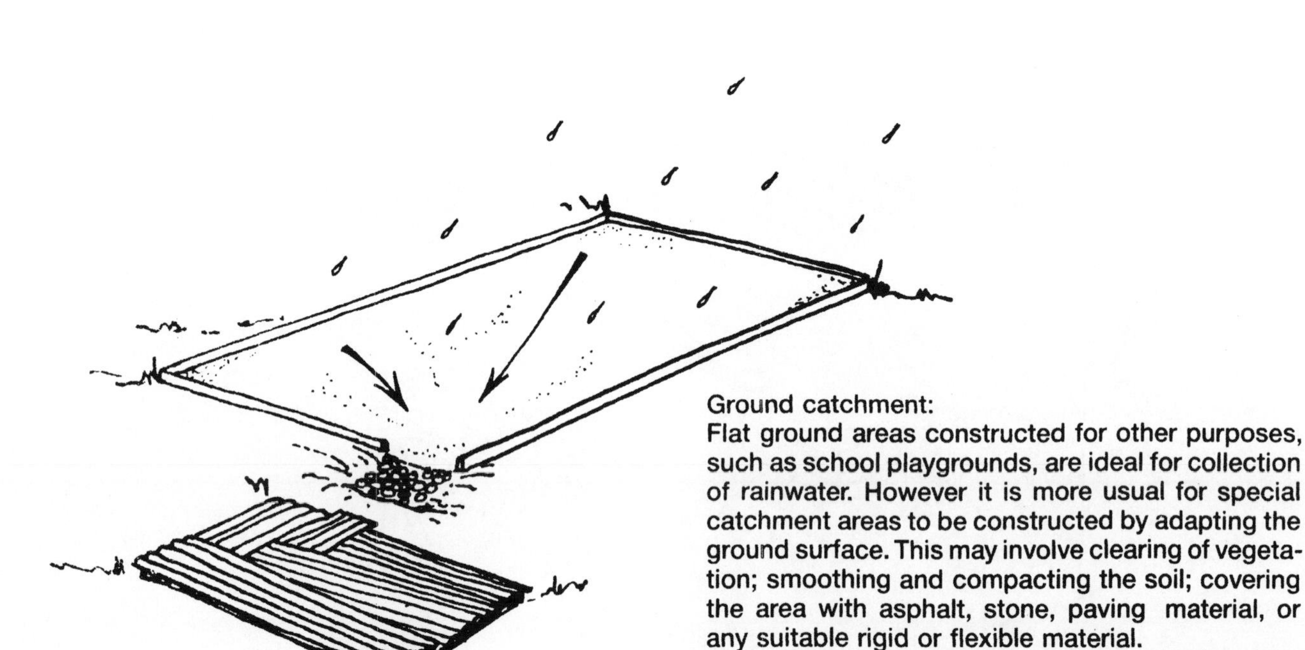

Ground catchment:
Flat ground areas constructed for other purposes, such as school playgrounds, are ideal for collection of rainwater. However it is more usual for special catchment areas to be constructed by adapting the ground surface. This may involve clearing of vegetation; smoothing and compacting the soil; covering the area with asphalt, stone, paving material, or any suitable rigid or flexible material.
Use of a ground catchment area can be much more expensive to construct and maintain than roof catchments. There may also be considerable pollution problems, and unless adequate treatment is available, water collected in this way should only be used for irrigation and cattle.

Collection

From roofs: This is usually done by means of a gutter and downpipe system. Gutters may be made from local wood or bamboo, or they can be preformed galvanised iron or PVC. The use of guttering may be a problem for some households, especially in low-income areas, but costs can be kept down if local materials are used.

Water must be carried from the eaves to the storage tank, and the simplest gutter and downpipe systems will be the cheapest to construct and maintain. The eaves gutter must be large enough to carry the run off expected during heavy rainfall. In general semi-circular gutters of about 200mm width will cope with all but the heaviest rainfall, although if local materials are used this may not be possible. The gutters must also be firmly supported or attached to a roof or wall, since they become very heavy when full of water. They should be fixed to slope slightly towards the storage tank so that no stagnant pools can form.

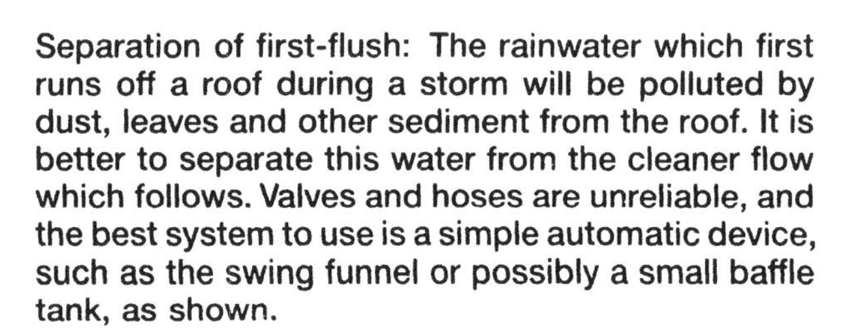

Separation of first-flush: The rainwater which first runs off a roof during a storm will be polluted by dust, leaves and other sediment from the roof. It is better to separate this water from the cleaner flow which follows. Valves and hoses are unreliable, and the best system to use is a simple automatic device, such as the swing funnel or possibly a small baffle tank, as shown.

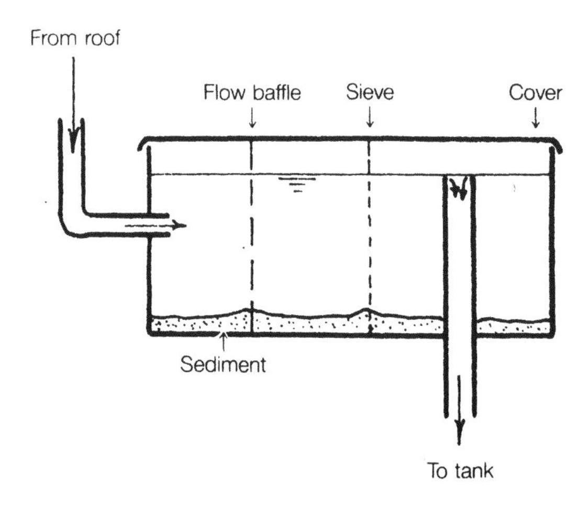

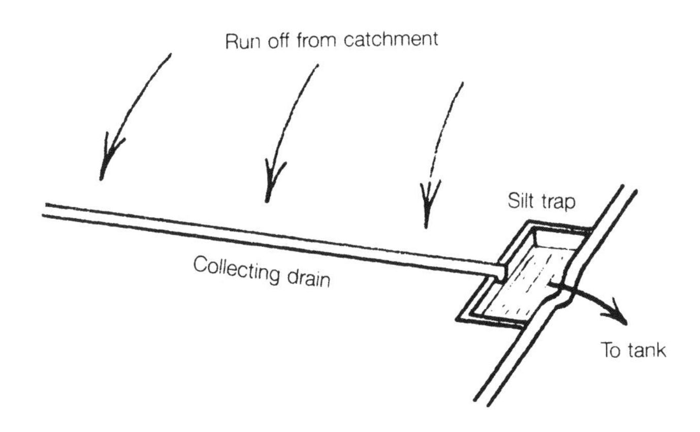

Collection from the ground: The slope of the catchment surface directs rainwater towards the storage tank, possibly by means of collecting drains. Water collected in this way is bound to be polluted, especially by sediments, and the best way to remove some of this material is by means of a silt trap.

RAINWATER HARVESTING

Storage

Rainwater can be stored in either underground or surface tanks of varying sizes. Whichever is used, it is necessary to completely enclose the tank and provide a tap or pump, in order to prevent pollution.

Underground storage: Large tanks can be built which are cool and easily protected, and which have almost no loss from evaporation. Also, by building underground, there is a considerable saving in space.
Materials used may vary from simple compacted earth (where ground conditions allow) and plastic sheets, to substantial reinforced concrete tanks.

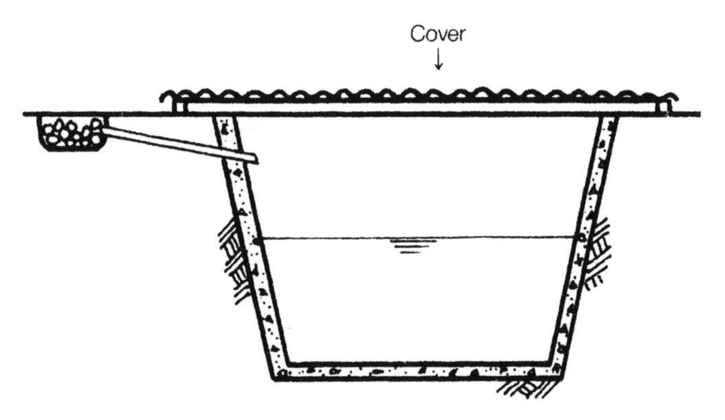

Surface storage: The size of storage vessel used will depend to some extent on the amount of rainfall and the size of the roof catchment involved. It will also depend on the economic circumstances of the householders and the availability of local materials. Buckets, barrels, clay pots and oil drums are all common. Locally produced cement mortar jars may have quite a large capacity. Unreinforced jars and bamboo-cement tanks are possible up to about 4.5m. Larger sized tanks are usually made from ferrocement, concrete or brick.

(Domestic storage tanks will be dealt with in detail in a later Technical Brief).

Water quality

Rainwater collected from roofs is usually much cleaner than that collected from the ground, and so is more suited to drinking and cooking purposes. It is however, still polluted to some extent by bird droppings, dust and leaves. Bird droppings can cause some bacterial contamination. Pollution can also occur during storage, if the tank is not properly covered or sealed. Organic matter will rot in the tank, and bacteria will multiply, and so it is important that a filter system or diversion of the first flush be used. In addition, the tanks should be kept cool and dark, and they should be regularly cleaned out.
Simple treatment of stored water might include boiling, pot chlorination or use of Moringa seeds as a coagulant.

For further information:

1. Berkovitch I. *Harvests of rain.*, Waterlines Vol. 1, No. 1, July 1982
2. Dian Desa. *Water purification with Moringa seeds*, Waterlines Vol 3, No 4, April 1985.
3. Pacey A. *Rainwater Harvesting*, Intermediate Technology Publications, 1986.
4. United Nations Environment Programme. Rain and Stormwater Harvesting in Rural Areas. 1983.

Susan Ball, WEDC, Loughborough University of Technology, UK.

12. Septic tanks and aquaprivies

Septic tanks and aquaprivies both provide primary treatment for wastewater, the only difference between them being the location of the tank in relation to the latrine, and the amount of effluent received by each.

They consist of sealed tanks buried in the ground and receiving household wastewater and excreta, which are partially treated within the tanks.

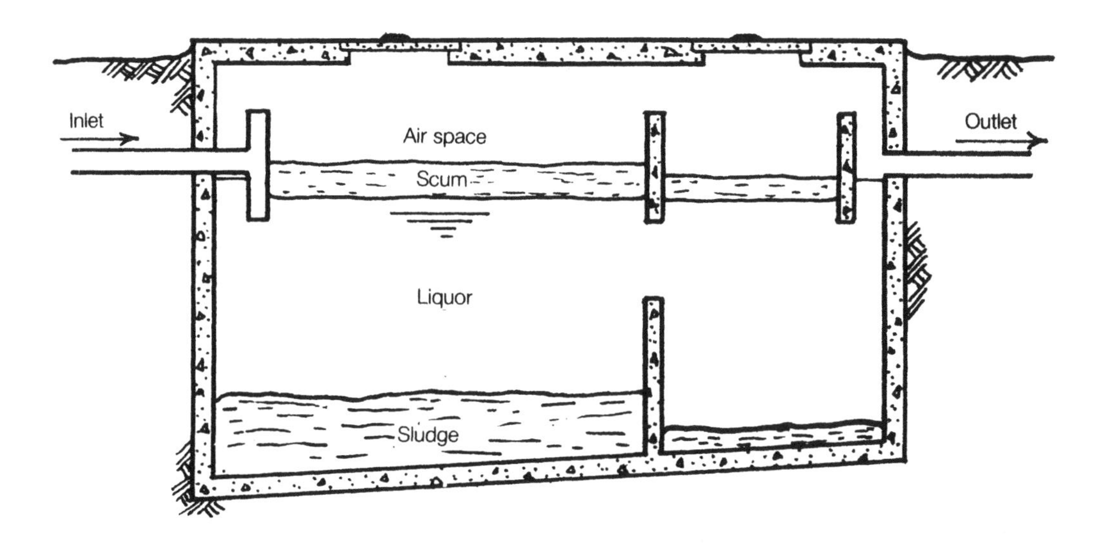

A septic tank is separated from the house and receives waterborne sewage, probably consisting of both toilet waste and household sullage.

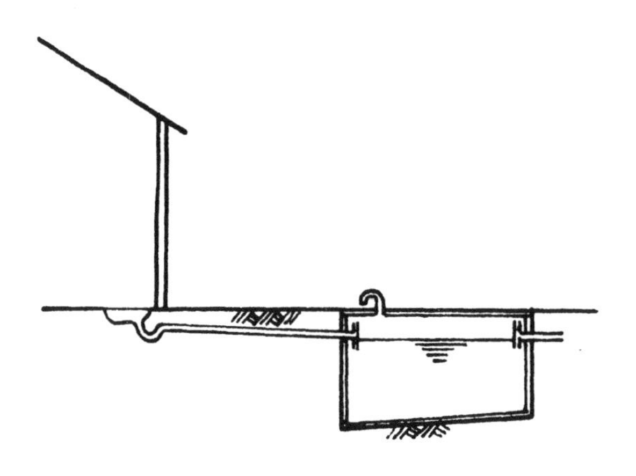

An aquaprivy is located directly below a latrine, and water is only needed to maintain the water level in the tank. The amount of wastewater treated will probably be less than for a septic tank, but is likely to be stronger.

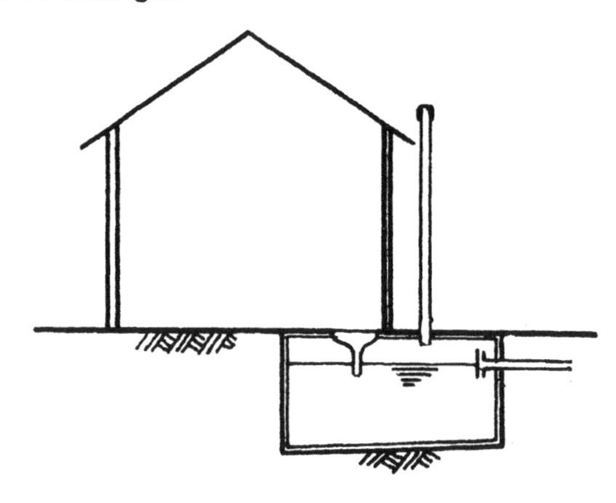

Either may be used to provide partial sewage treatment for individual houses or communal latrines, but they are a comparatively expensive form of sanitation. They are best suited to medium- or low-density housing areas with large plot sizes. Problems arise when they are used in high-density housing areas, and consideration of an adequate means of disposal of the effluent is often neglected.

3. Tank capacity

Tanks should be constructed to have a volume sufficient to allow for the estimated quantities of sludge and scum that will accumulate between desludging operations, and to accommodate the expected wastewater flows for a 24-hour period.

The total liquid storage capacity can be calculated in separate stages.

i) Liquid capacity, $V = Pq$

Where P = number of people expected to use the tank, and q = average daily sewage flow per person.

Approximate flows are:

Septic tank (WC only)	15-40 lcd
Septic tank (full plumbing)	50-120 lcd
Aquaprivy (WC only)	5-18 lcd

ii) Sludge and scum capacity, $W = Pnfs$

Where P = number of people expected to use the tank,

n = number of years betwen desludging operations,

f = a factor relating to the rate at which the sludge is digested, and

s = the rate at which sludge and scum accumulate.

iii) Total liquid capacity, $C = V + W$, or $C = (1.5)W$, whichever is the greater. It is better to build a tank larger than the minimum size because future needs cannot always be predicted.

Interval between Desludging operations. (Years)	Ambient Temperature		
	More than 20°C throughout the year.	More than 10°C throughout the year.	Less than 10°C during the winter months.
1	1.3	1.5	2.5
2	1.0	1.15	1.5
3	1.0	1.0	1.27
4	1.0	1.0	1.15
5	1.0	1.0	1.06
6 or more	1.0	1.0	1.0

Values of 'f' for calculation of sludge and scum capacity.

Material used for anal cleansing.	Flush toilet or latrine wastes only.	Both Domestic sullage and toilet wastes.
Water, Soft paper.	25	40
Leaves, Hard paper.	40	55
Sand, Stones, Earth.	55	70

Values of 's', in litres, for calculation of sludge and scum capacity.

Design details

1. Provide a pipe, at either the tank or latrine, to vent waste gases.

2. Ensure the wastewater inlet is below liquid level.

3. Provide a barrier at the outlet pipe to avoid discharge of sludge or scum.

4. Provide access for maintenance and desludging.

5. A slight slope on the tank base makes desludging easier.

6. Tanks having more than one compartment should have openings through dividing walls at liquor level.

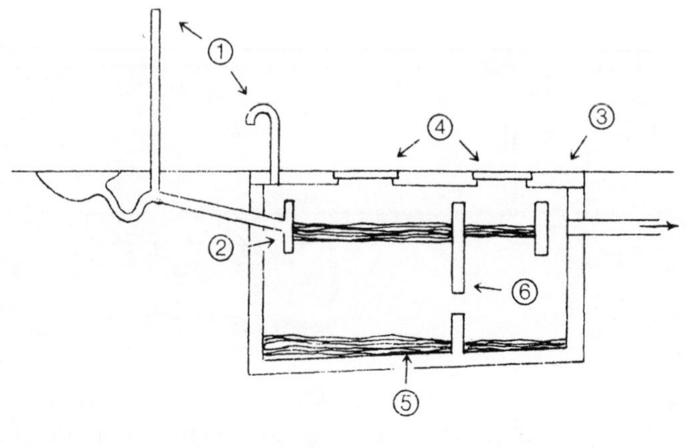

7. Never attempt to disinfect or completely empty tanks. New tanks will start to work quickly if a bucket of sludge from a working tank is poured into them.

Processes within the tank

1. Separation of solids
Dense material settles to the bottom of the tank where it undergoes digestion to form sludge. Light material rises, floating to the surface to form scum. In between sludge and scum, is a liquid known as liquor, which contains organic and inorganic material in solution and fine particles in suspension which may coagulate and settle out.

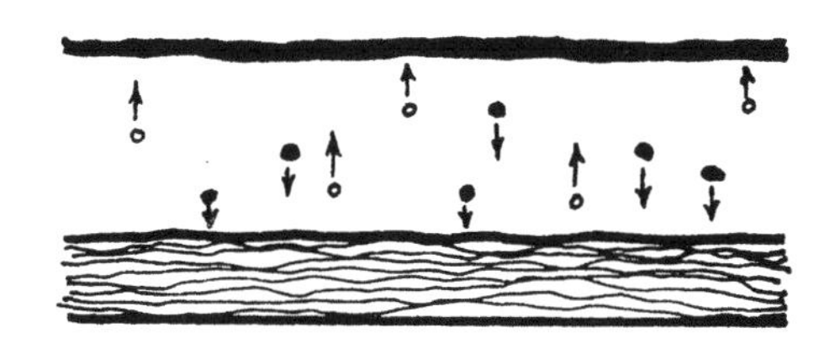

2. Digestion of scum and sludge
These are partially digested anaerobically by bacteria. The waste products, water, methane and carbon dioxide are produced. This causes a reduction in the amount of sludge which forms. Sludge must be periodically removed.

3. Stabilization of the liquor
Organic material remaining in the liquor is partially broken down by anaerobic bacteria.

4. Consolidation of sludge
As material settles to the bottom of the tank, its weight compacts lower layers of sludge. This thickens the sludge and reduces its water content.

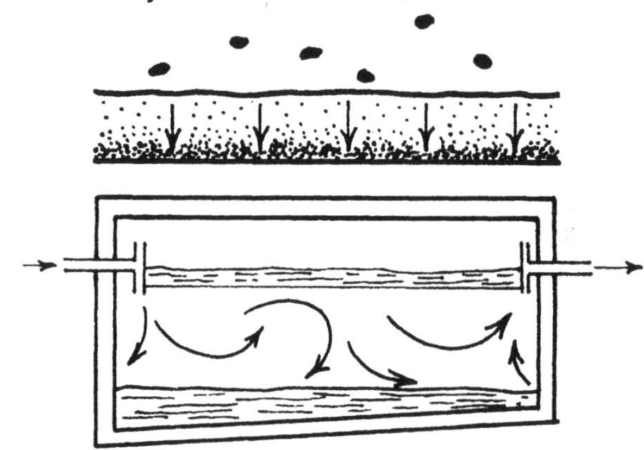

5. Mixing of the tank contents
The efficiency of the treatment process is reduced if the tank contents are disturbed, but some mixing is inevitable because of gas production and flow of liquid through the tank.

6. Production of micro-organisms
The tank forms a hostile environment for many micro-organisms, but the effluent will contain large numbers of bacteria, viruses and other potentially harmful organisms.

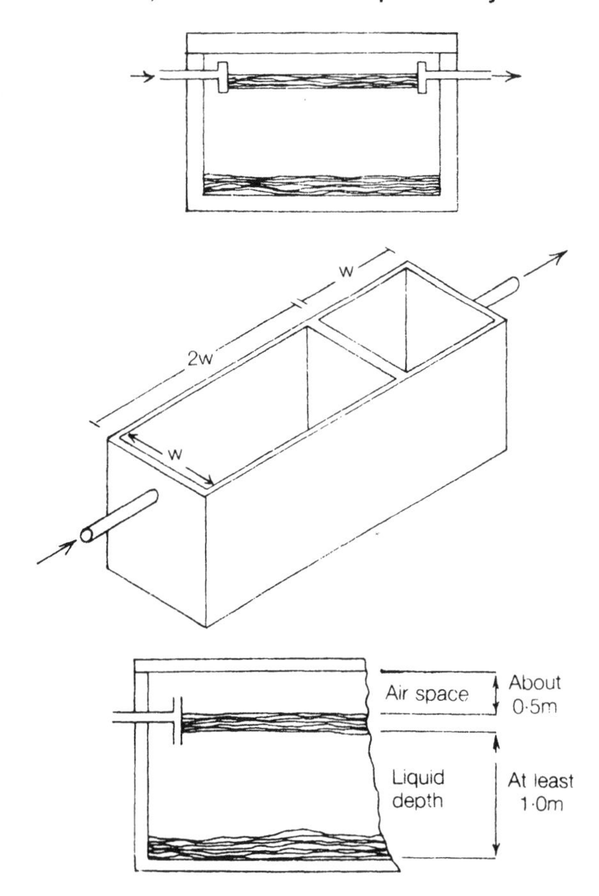

Design considerations

1. Number of compartments
Tanks having a single compartment can operate well, but there is a danger of sludge being disturbed and discharging through the outlet pipe.

Tanks with two compartments are more usual. The first compartment should have a length equal to about twice its width, and the second compartment should have a length equal to its width. Such a design reduces the likelihood of disturbed sludge being carried out of the tank in the effluent.

Tanks with three or more compartments have been suggested but there is little evidence of improved performance resulting from this.

2. Depth
Depth of liquor in a tank should be at least 1.0m, but a depth of 1.5m or more is preferable. There should be an air space of about 0.5m between the liquor and the tank roof.

SEPTIC TANKS AND AQUAPRIVIES

Disposal of effluent from a tank

Careful consideration should be given to the selection of a suitable means of disposal of tank efluent. Possible options are sewers, soak-aways and evapo-transpiration beds.

Soak-aways
Effluent is allowed to filter into the ground. Large areas of land may be needed, and the size of trenches or pits should be estimated by performing soil percolation tests.

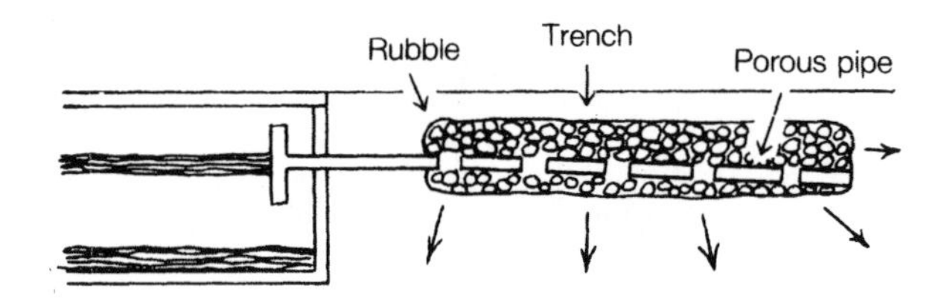

Infiltration rates for porous soils are likely to be in the range 10 – 30 litres per day per square metre of sidewall.

If a single pit or trench is used, periodic emptying will be necessary. It is better to provide two pits or trenches which can be used alternatively, allowing the resting pit or trench to recover between periods of use.

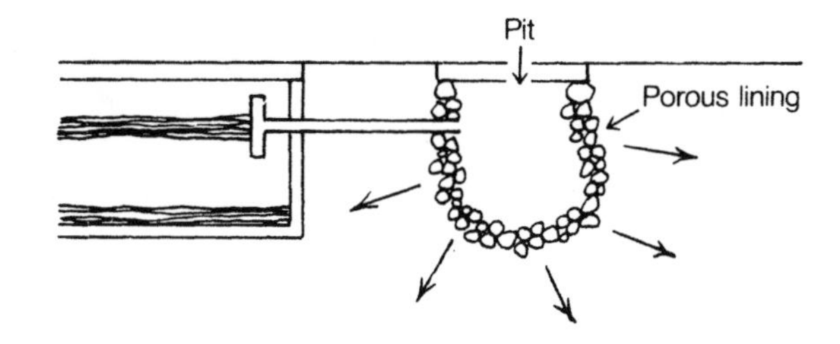

Sewers
These may be either conventional or small bore, but some authorities may prohibit the connection of septic tanks to sewerage networks because of the septicity of the tank effluents.

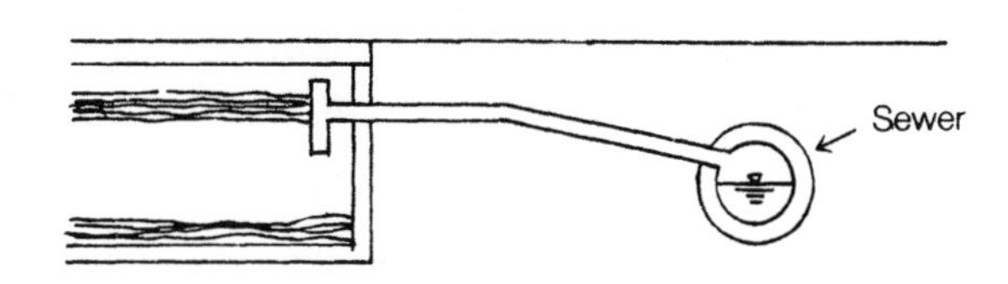

Evapo-transpiration beds
Effluent is evaporated from areas of land on which a steadily growing crop is planted. Large areas of land may be required.

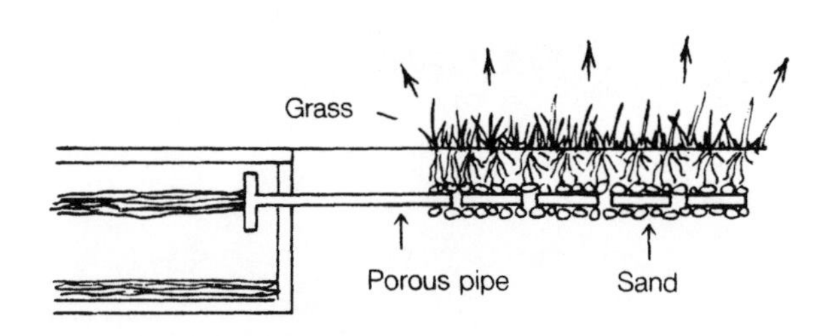

For further reading:

1. Cairncross, S. and Feachem, R. G., *Environmental engineering in the Tropics.* John Wiley and Sons, 1983.

2. Feachem, R. G., et al., *Health aspects of excreta and sullage management*, World Bank, 1981.

3. Mara, D., *Sewage treatment in hot climates*, John Wiley and Sons, 1983.

4. Pickford, J., *The design of septic tanks and aquaprivies*, Overseas Building Note No. 187, September 1980, Building Research Establishment, Watford, England.

Michael Smith, WEDC, Loughborough University of Technology, UK.

13. Handpumps

Shallow pumps

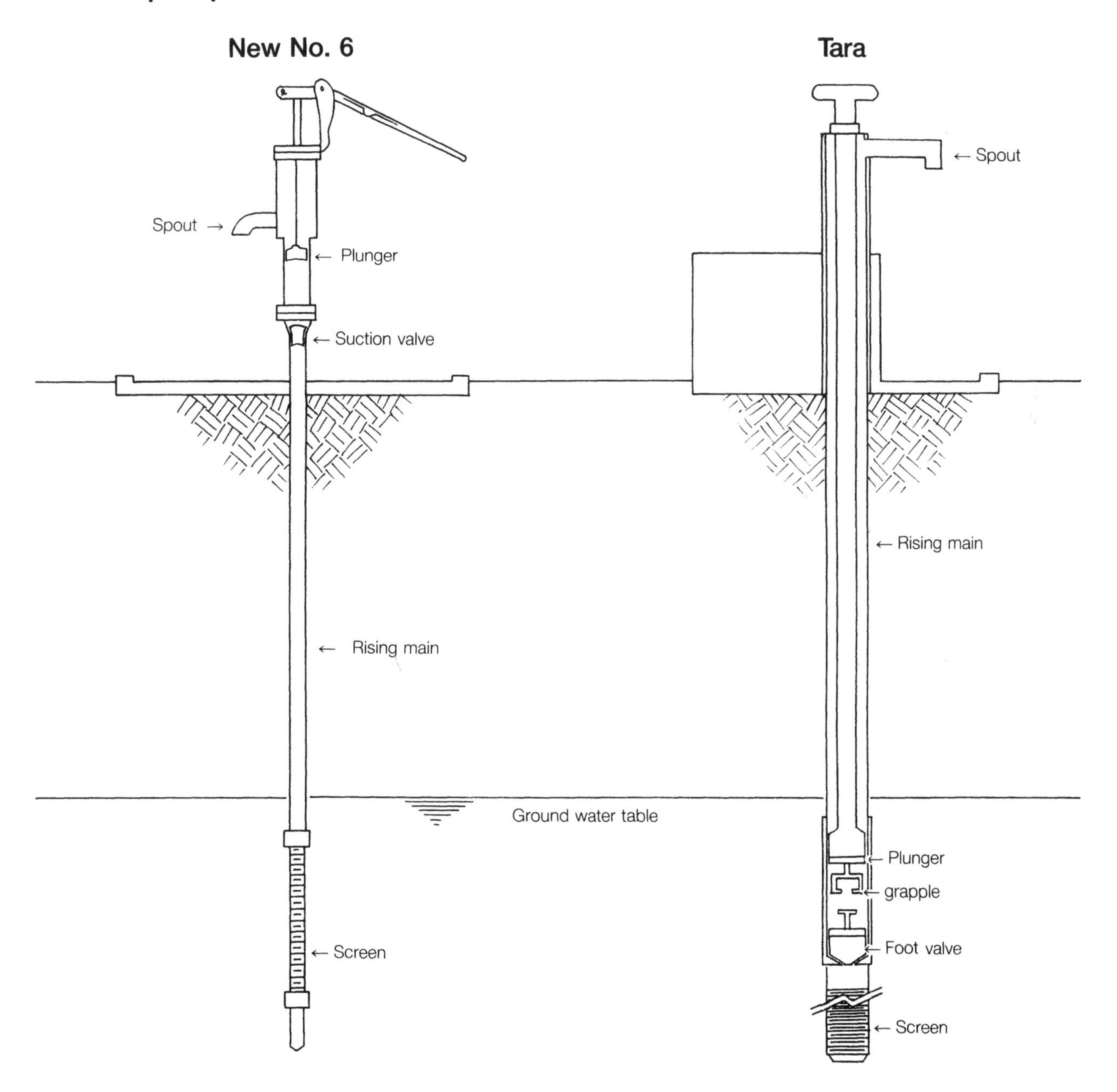

The most common form of shallow pump works by suction. The pumping mechanism is contained in the pump head above ground level. Atmospheric pressure limits maximum pumping depth to 7m at sea level (or 6m at 1,100m elevation). This form of pump requires priming to begin operation. This may mean that polluted water can be introduced into the well.

A direct action pump such as the Tara does not have a lever handle and bearing. The pumping mechanism is moved to a cylinder below the level of the groundwater and when the handle is raised the plunger lifts the column of water to the surface whilst the foot valve opens to allow the cylinder to refill. The pump is not limited by atmospheric pressure, though it is most efficient at depths of up to 15m. The pump rod is a hollow sealed PVC pipe, which, by using the effect of bouyancy, distributes the force required for pumping more equally between the up and down stroke.

Deep Pumps

For many low-income communities, the installation of a handpump is the cheapest and most effective means of providing an improved water supply. Deep pumps, where the forces on the components are high, must be designed for use by many different people for up to 18 hours every day.

Pumps should be maintained by local people without special tools and lifting equipment. This is known as a VLOM capability (Village Level Operation and Maintenance). There are many hundreds of different types of handpumps and manufacturers, and the few illustrated here have been chosen to show just some of the alternatives.

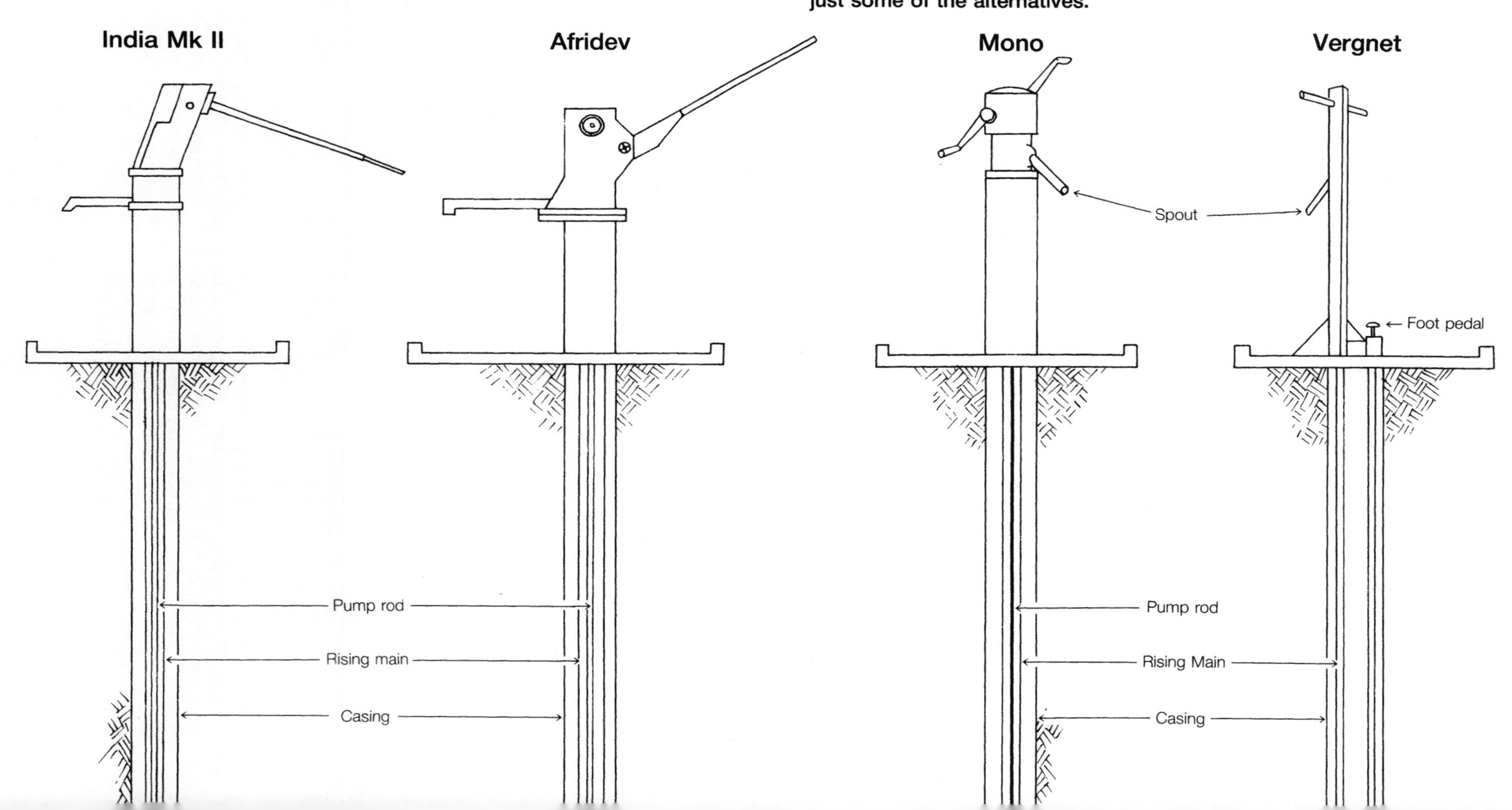

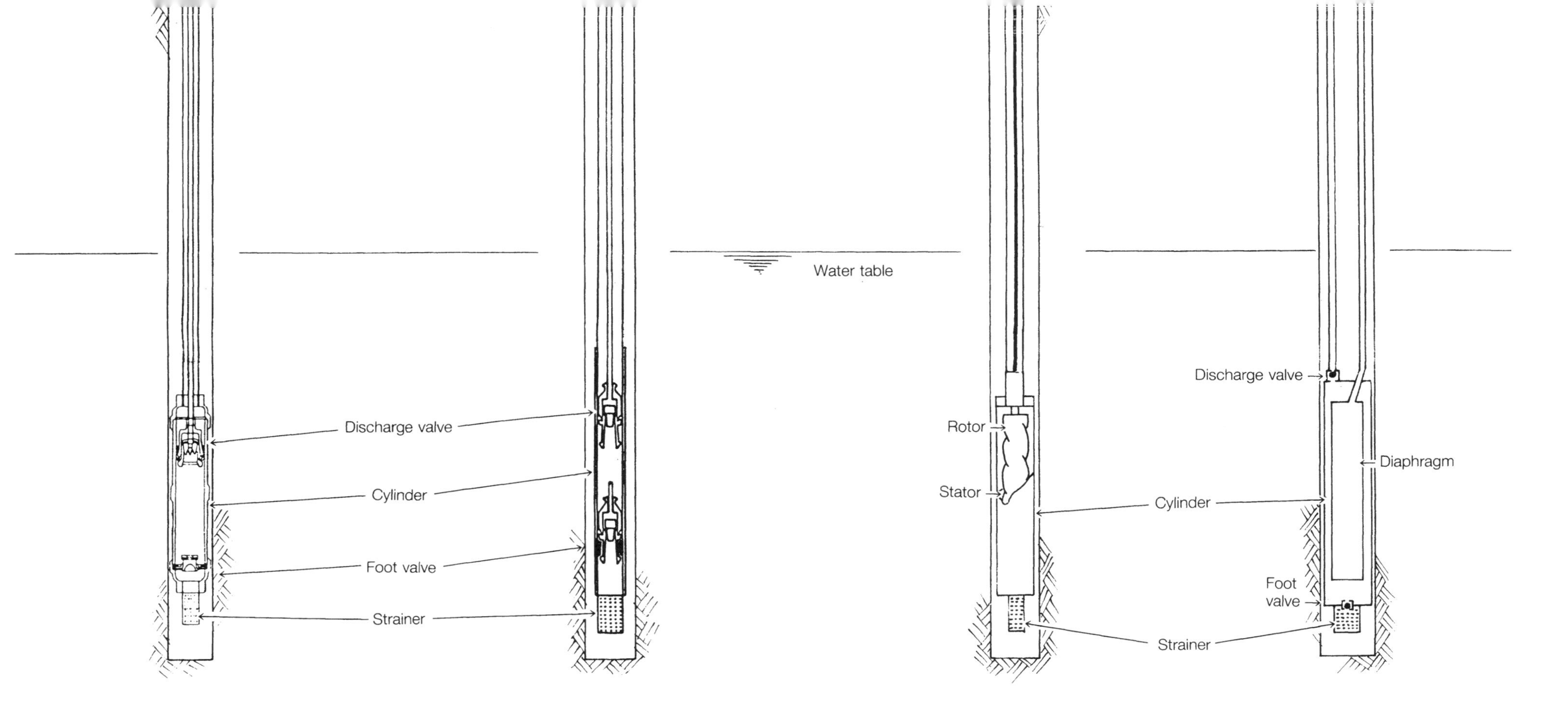

Positive displacement pumps

1. Lifting and lowering the pump handle of the India Mk II produces vertical displacement of the pump rod.
2. The discharge valve (plunger) attached to the lower end of the pump rod closes as it moves up, thereby lifting water and allowing the foot valve (check valve) to open and refill the cylinder.
3. The foot valve then closes as the discharge valve opens on the down stroke, moving through the water without pumping.

The Afridev works in the same way as the Mk II, but requires only a single tool for all maintenance and repair. The pump rods have a mechanical linkage rather than screwed connectors. The discharge valve and foot valve (made from identical plastic parts) can be withdrawn for maintenance without having to remove the cylinder and rising main. A similarly designed cylinder is now being developed for the Mk II.

Progressive cavity

1. Turning the handles of the Mono rotates the pump rod by means of a gear box in the pump head.
2. A foot valve is not needed. A stainless steel rotor twisting within the rubber stator in the pump cylinder creates a moving 'progressive cavity' which 'screws' the water upwards. This design can easily be motorized as finance becomes available.

Diaphragm pump

1. Pressure on the Vergnet foot pedal forces water down the closed pipe into the diaphragm or elastic sleeve, which then expands.
2. The increasing volume of the diaphragm forces water out of the cylinder, up the rising main, and out of the spout.
3. As pressure is taken off the pedal, the diaphragm contracts and water enters the cylinder through the foot valve.

HANDPUMPS

Every handpump must have a concrete surround to prevent polluted water seeping down the side of the casing and polluting the borehole water. This is also needed so that people drawing water do not have to walk through mud or stagnant water where they may pick up disease.

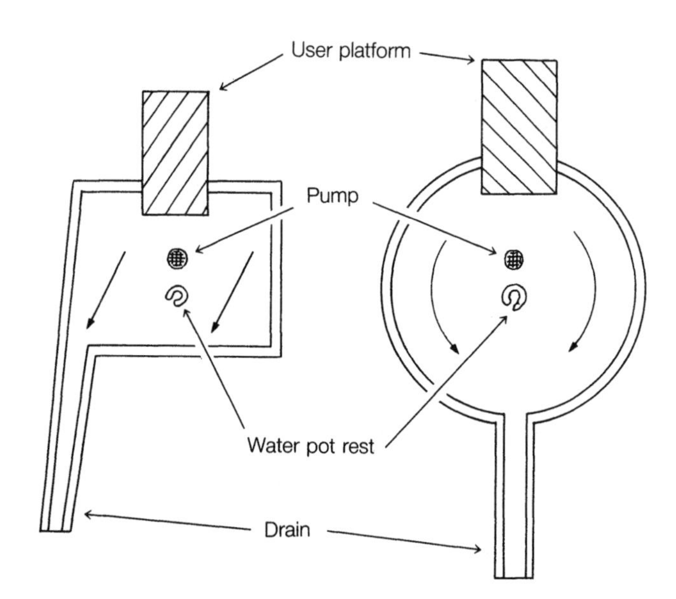

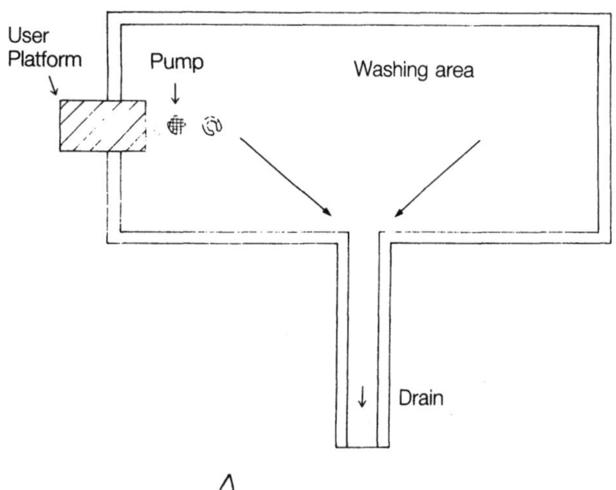

Platforms can be of all shapes and sizes as long as there is at least 600mm of impervious material around the spout. Concrete is normally used to make the surround. It is important not to over-economize in the design of the slab – especially its thickness – as a cracked slab is worthless.

There is usually a considerable amount of waste-water from a handpump. This must be disposed of into a natural drain or a soakage pit, or perhaps a kitchen garden.

For further information:

Kennedy, W.K. and Rogers, T.A., *Human and Animal-powered Water Lifting Devices*, I.T. Publications, 1985.

World Bank/Rural Water Supply Handpumps Project, *Technical Papers 6, 19, 29*, and *Community Water Supply — the Handpump option*, World Bank, 1986.

Richard Franceys, Water, Engineering and Development Centre, Loughborough University of Technology, UK.

14. Above-ground rainwater storage

A rainwater collection system has three essential components.

- an impervious roof to collect rainfall;
- gutters and downpipes to convey collected water to a storage tank;
- a tank for the storage of collected rainfall.

The tank is the most expensive item.

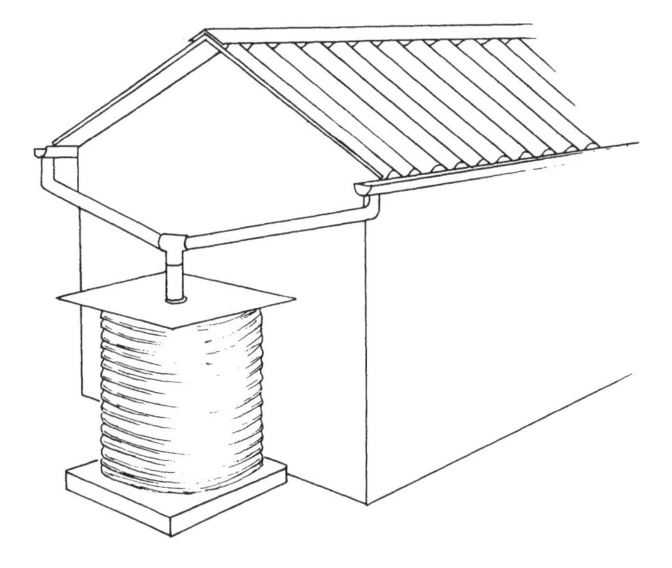

The capacity of the tank

What is needed:
The total quantity required each day multiplied by the longest period without rain.

What is available:
The total quantity of rain falling on the roof during the rainy season, less the water drawn from the tank during that period, less losses due to evaporation and soaking into the roof etc.

A cumulative rainfall diagram may be used to calculate the required capacity.

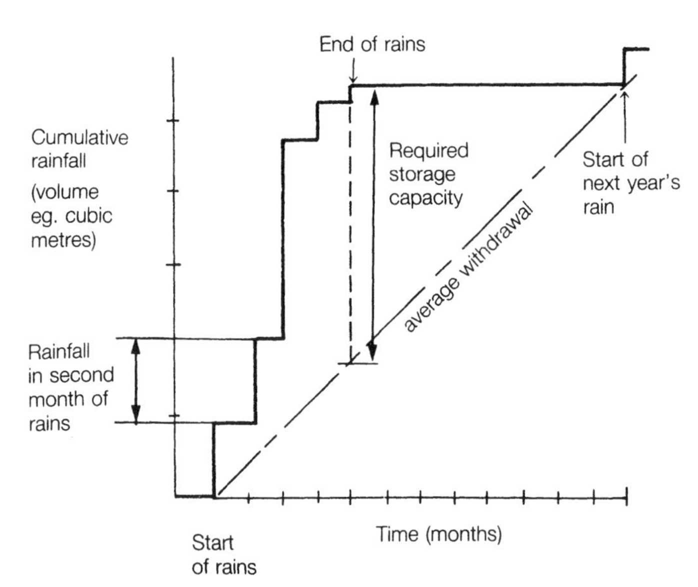

Materials which are used for tanks include:

- corrugated iron
- cement jars and cement-covered baskets
- granary bins lined with mortar
- concrete or sancrete blocks, bricks and masonry
- pre-cast concrete rings and panels
- in-situ concrete and soil-cement
- ferrocement

ABOVE-GROUND RAINWATER STORAGE

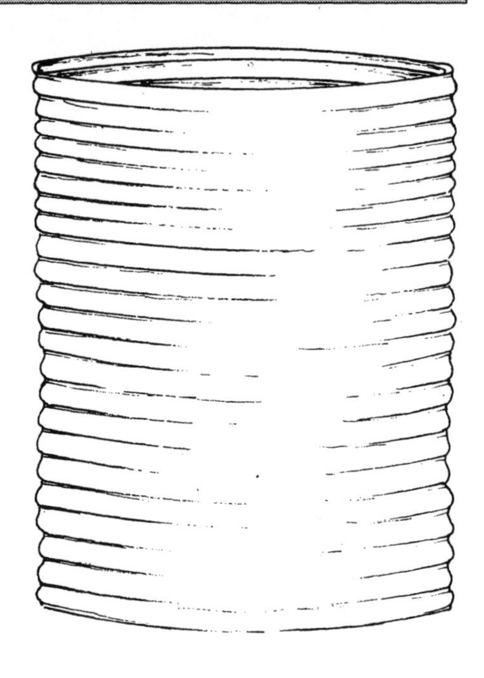

Corrugated iron

These are often available commercially.
Erection is simple on a flat concrete slab.
They are usually expensive.
Corrosion may set in quickly, especially at joints, so the life of the tank is often limited to 5-10 years.
Corrosion can be reduced by painting tanks inside and outside with bitumen.

Cement jars and cement covered baskets

Jars are made by spreading layers of cement mortar to a thickness of 10mm or more.

A sack is filled with straw, sand or sawdust, and a metal ring is placed at the top for the opening. (see Technical Brief No 1, Waterlines, Vol 3, No 1, July 1984.)

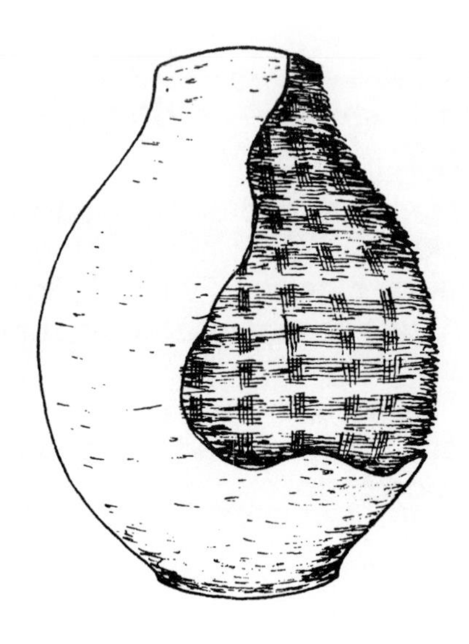

OR

A granary basket is made with woven sticks.
The capacity can be up to 3.5 cubic metres with plain mortar, or up to 10 cubic metres with reinforced mortar.

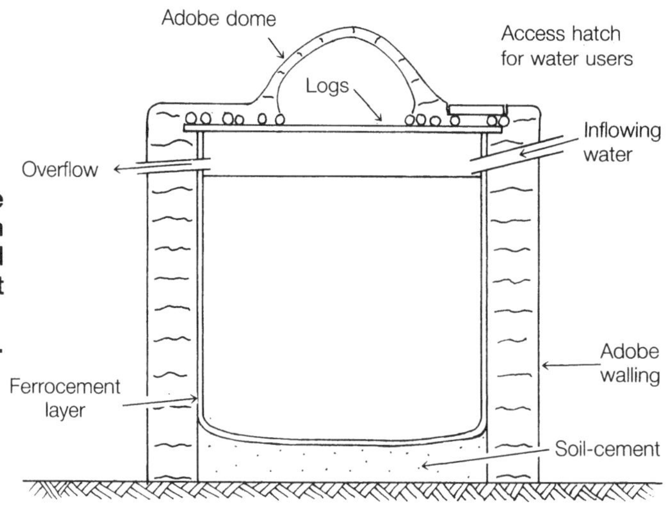

Granary bins lined with mortar

In Mali, bins built of adobe for grain storage have been converted to water tanks by lining with cement mortar or ferrocement 10mm thick. A curved base is made of soil-cement, (10 parts soil to 1 part cement).
The size is typically 2.6m diameter and 2.4m high.

Blocks, bricks and masonry

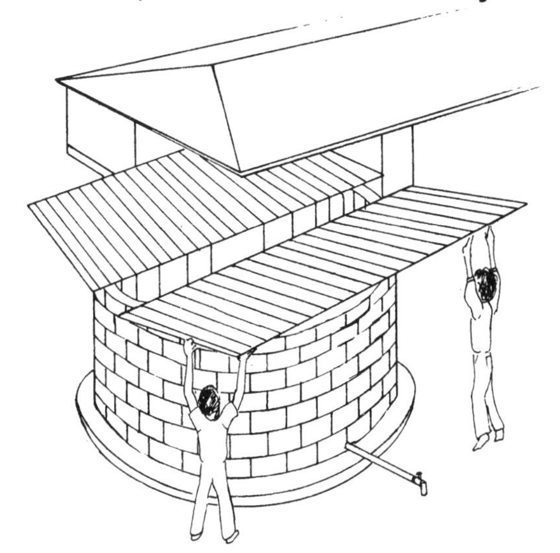

Depending on local traditional building practice, circular tanks are made with blocks of concrete or sandcrete; sun-dried, kiln-dried and adobe bricks; or masonry (natural stone). The inside is always plastered with cement mortar. The outside may also be plastered.

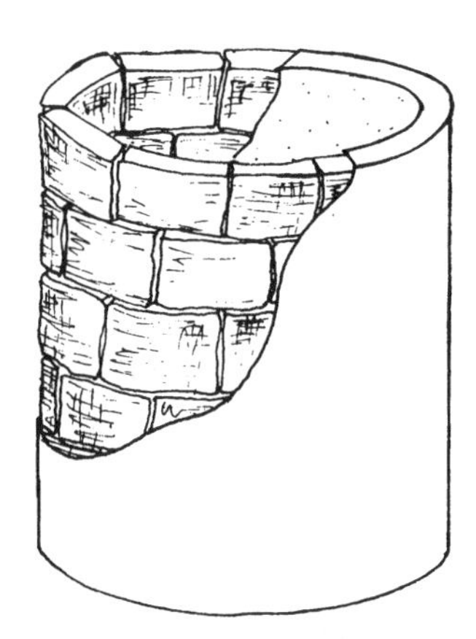

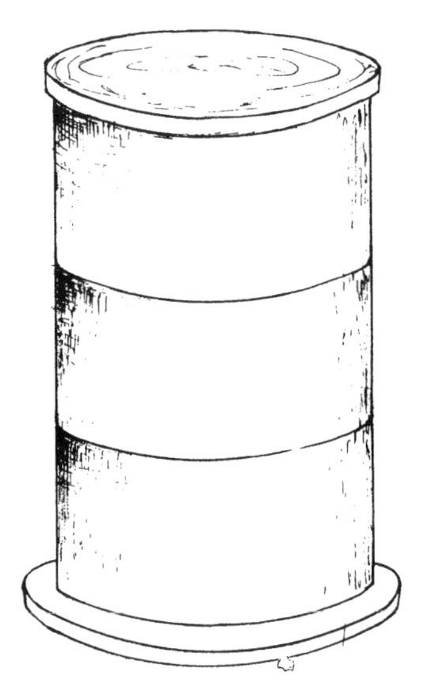

Pre-cast concrete rings and panels

Large diameter concrete pipes and rings fabricated for sewer manholes and septic tanks are built into a tank. Joints must be laid with a strong mortar mix.

Pre-cast concrete panels have been used in the South Pacific – first in the Cook Islands; also in Tonga, Kiribati, the Solomon Islands, Samoa, Papua New Guinea and Fiji.

Panels are 1.65m by 1.65m, reinforced with wires in both directions, or 1.20m by 1.20m unreinforced concrete, with wires sticking out for jointing.

A 1:2 cement:sand mix is used, reinforced with steel fibres.

Multiple tanks are made by joining panels.

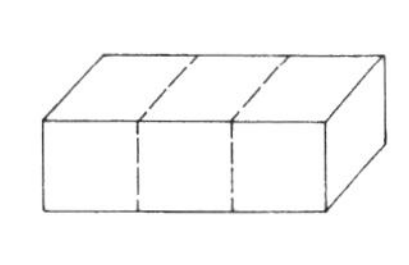

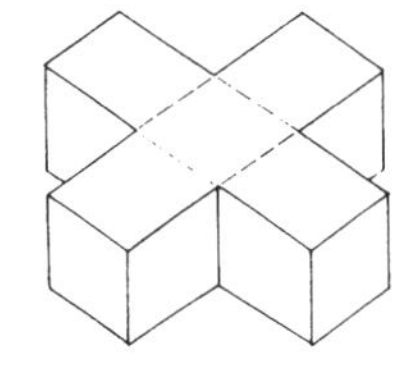

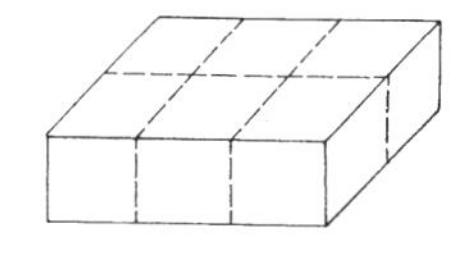

ABOVE-GROUND RAINWATER STORAGE

In-situ concrete and soil-cement

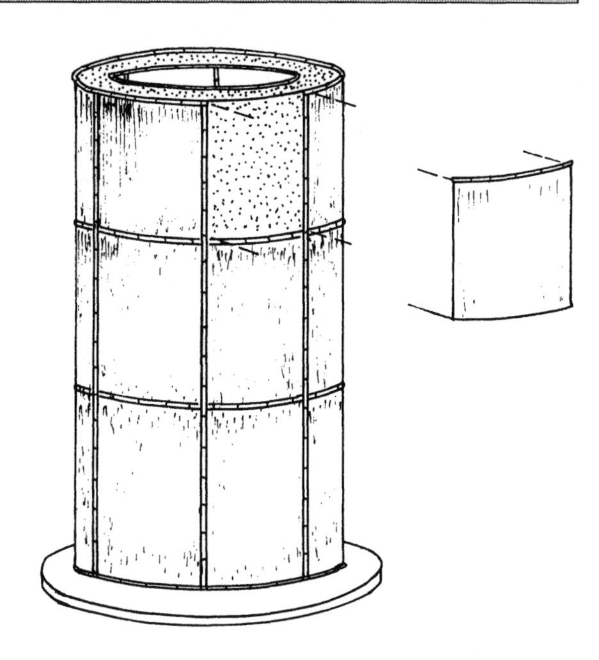

Circular concrete tanks are made with steel or plywood shutters. Steel shutters are expensive, but may be used many times.

Walls of 12:1 soil:cement, 100mm thick have been built in Brazil, using plywood shuttering. The tanks are rectangular.

(see Waterlines, Vol 5 No 1, July 1986, pages 25-28).

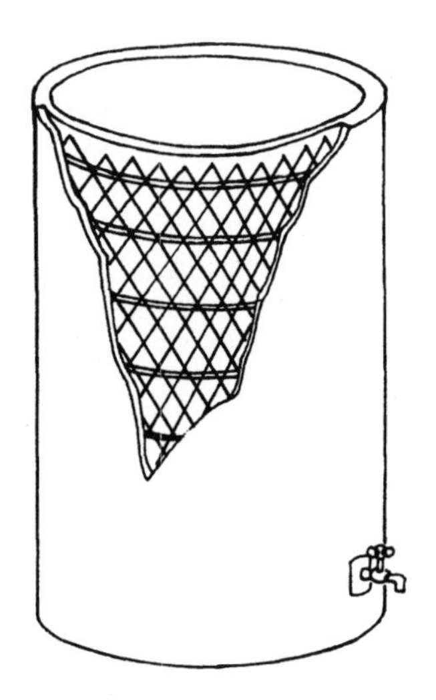

Ferrocement

Tanks with a capacity of 10 cubic metres have been built from ferrocement.

For domestic tanks a capacity of 10 cubic metres is provided by making the diameter 2.5m and the depth 2m.

Shuttering can be made from 16 sheets of corrugated iron.

Timber and adobe can also be used for shuttering.

The cement mortar is usually made with 1 part cement to 3 parts sand by volume, ½ part water to 1 part cement by weight.

The mortar is applied in layers not more than 10mm thick to give a total thickness of 40mm.

Reinforcment is provided by chicken wire, barbed wire or spot-welded wire mesh.

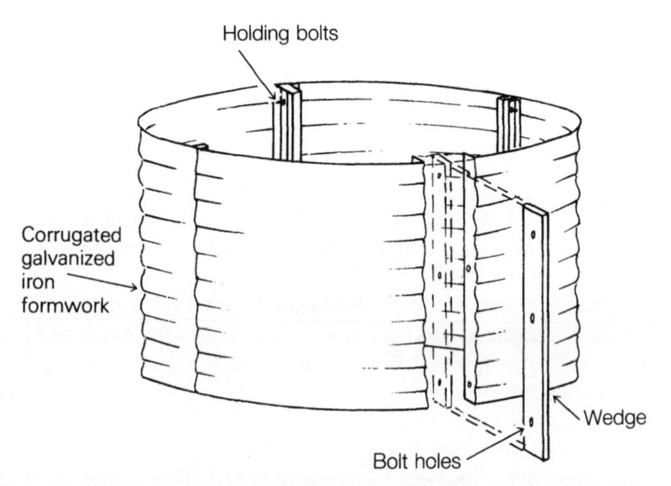

For small tanks the floor and walls are often made as a continuous construction, avoiding the need for a flexible joint at the base of the wall. The internal angle should be coved with extra mortar.

Alternatively, the floor can be made of puddled clay or a plastic sheet between layers of sand.

John Pickford, WEDC, Loughborough University of Technology, UK.

15. Slow sand filter design

Sand Filter Design

Slow sand filtration is a simple and effective technique for purifying surface water. It will remove practically all the turbidity from water, together with virtually all harmful eggs, protozoa, bacteria and viruses without the addition of chemicals and may frequently be constructed largely with local materials.

A slow sand filter consists basically of three different layers within a filter-box. These layers are from bottom to top: the underdrainage system, the gravel layer and the sand. It is only the sand which plays any part in the treatment process.

The underdrainage may consist of:

* perforated pipes of asbestos cement
* porous or perforated unglazed pipes
* perforated pipes of PVC

or of

* concrete tiles
* household bricks
* large gravel (40 - 100mm)

If pipes are employed, a series of lateral drains (80mm diameter) are connected to a main drain.

Perforations of 2 to 4mm diameter are made on the underside of the lateral drains at intervals of 150mm.
Cross-sectional area of main drain = sum of cross-sectional areas of all lateral drains.

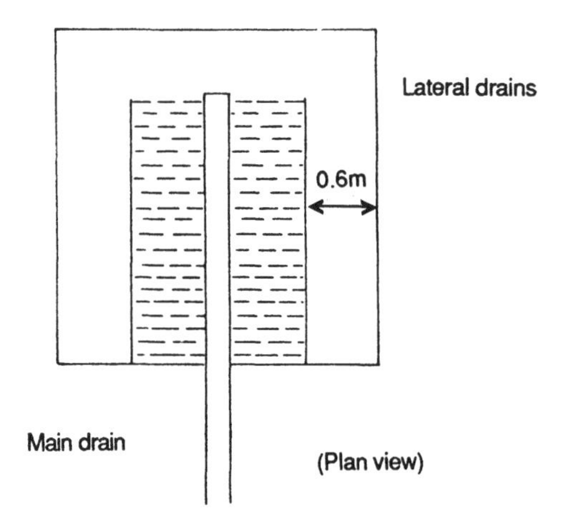

Standard bricks.

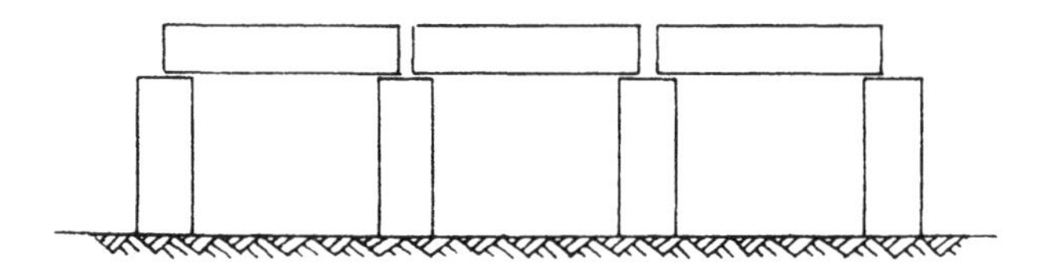

Concrete tiles - whole tiles are set on quarter-tiles as illustrated.

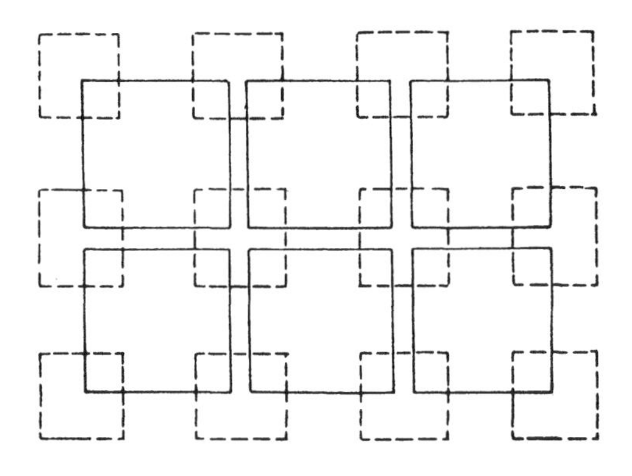

Large gravel (40 - 100mm).

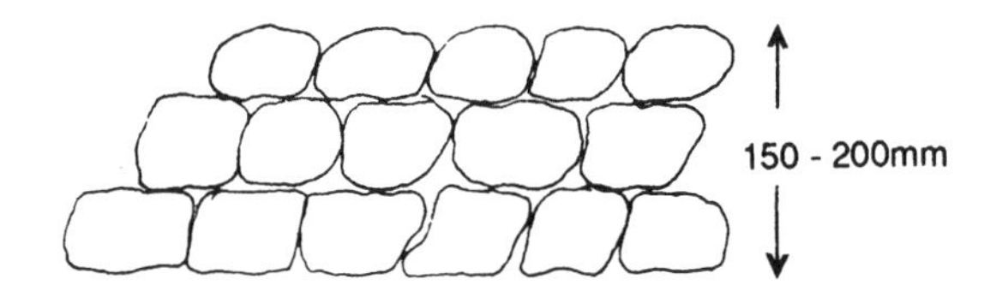

The underdrainage should not be closer to the wall than 0.6m.

SLOW SAND FILTER DESIGN

The Gravel Layer

The gravel layer is arranged in four graded levels. All gravel must be clean.

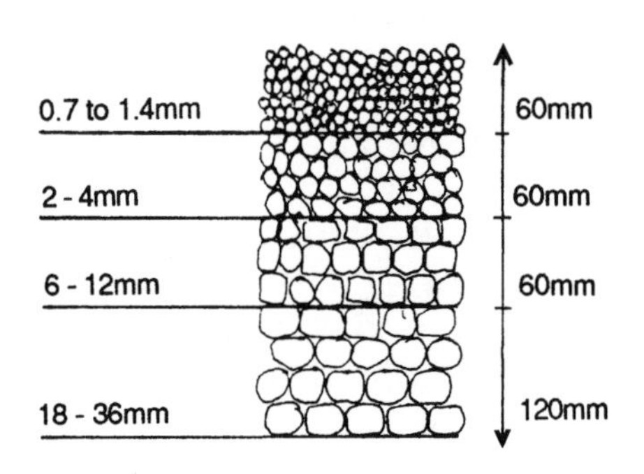

The gravel layer should not be closer than 0.6m to the walls. This means that any water which runs quickly down the walls and does not filter through the sand layers (ie, it 'short-circuits' the system) must pass through some depth of sand before entering the gravel and underdrainage.

The Sand Layer

Is characterized by:

Effective Size (E.S.) - mesh diameter (mm) of a sieve which retains 90 per cent of the sand.

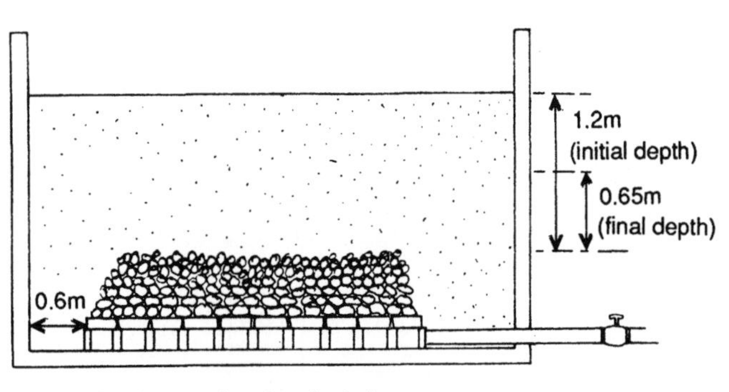

Sand, gravel and underdrainage

Uniformity Coefficient (U.C.) - mesh diameter (mm) of a sieve which retains 60 per cent of the sand, divided by the effective size.

E.S. - between 0.2mm and 0.4mm

U.C. - less than 3.0, preferably less than 2.0

Suitable sand is usually easy to find locally. If any grading is necessary, it is normally sufficient to remove only the coarsest grains and the very finest grains.

Filter Box

The filter box may be constructed either with vertical sides or with sloping sides.

If designed to have vertical sides it may be constructed of either:

* mass-concrete
* ferrocement
* masonry
* reinforced concrete

During construction the wall must be roughened where it will be in contact with the sand in order to prevent short-curculting (see section on the gravel layer).

If the filter box is designed to have sloping sides it may be constructed either of:

* mass-concrete
* masonry
* puddled clay
* rip-rap

Commonly, masonry will be employed and the system made watertight by adding a layer of puddled clay

Sloping walls are simpler to construct and can usually be made with locally available material but cannot be guaranteed to be always watertight.

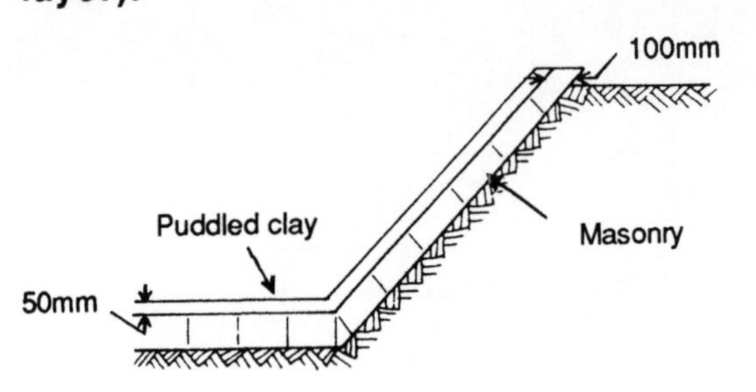

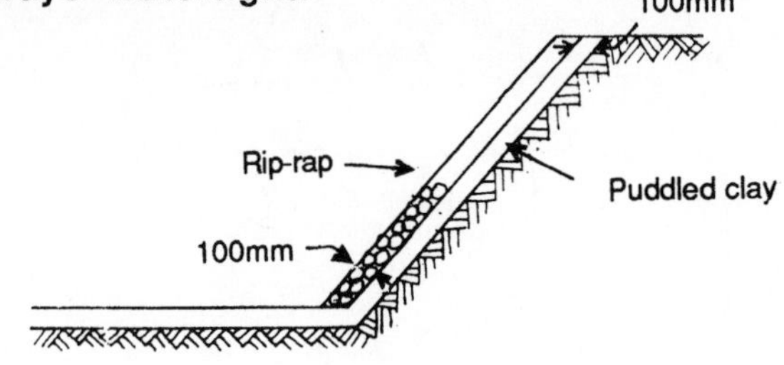

Inlet Arrangements

There are two main types of inlet arrangement.

The concrete tile is a splash tile to prevent water falling directly onto the sand-bed and eroding it.

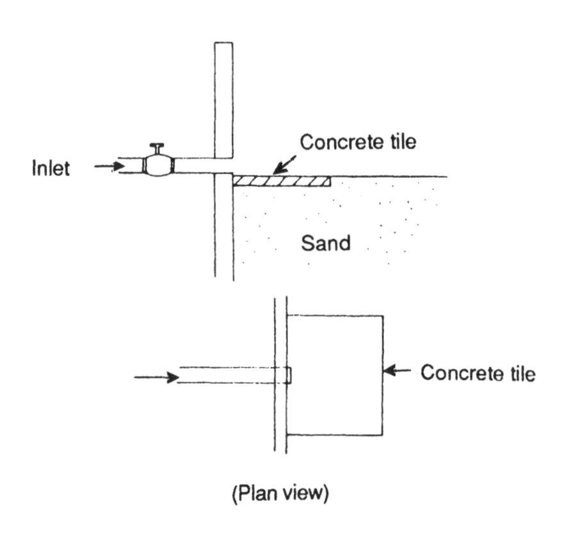

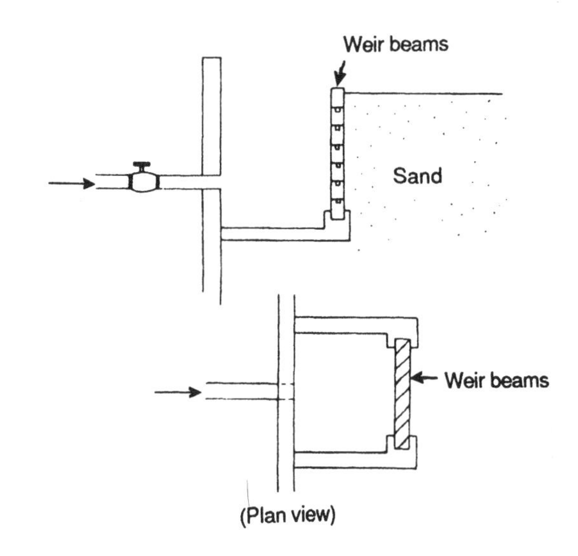

The inlet is controlled either by a hand-operated gate valve or by a float-controlled butterfly valve.

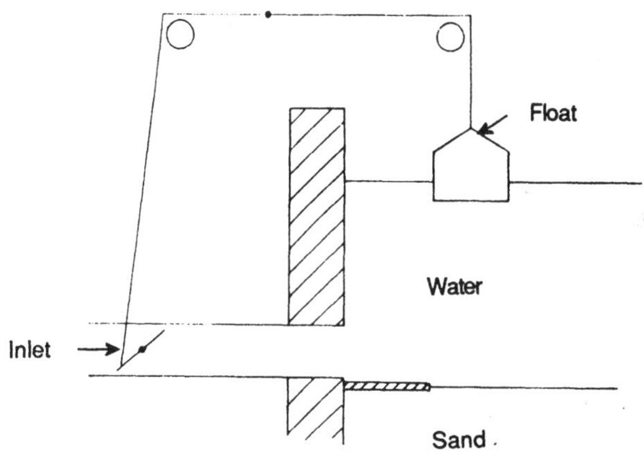

Water Reservoir

This is normally maintained at a constant depth of between 1.0m and 1.5m.

Outlet Arrangement

The outlet flow is maintained at the design flow rate by a hand-operated gate valve which is adjusted every day. It is essential to provide an outlet weir which is above the height of the sand bed.

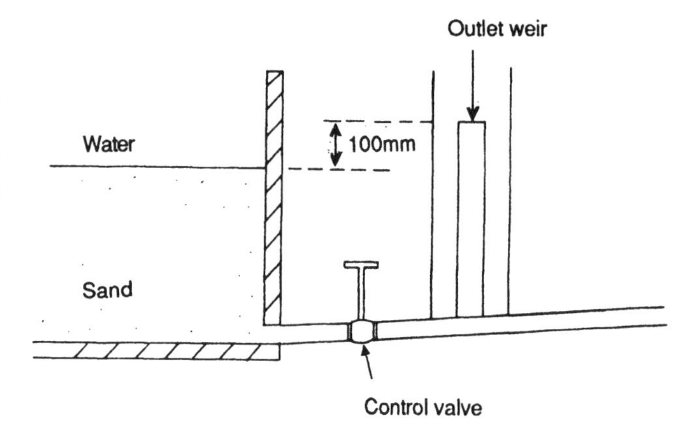

SLOW SAND FILTER DESIGN

The size of the filter

The size of a slow sand filter is determined by several factors.

For example:

Population of 1000 with water consumption of 100 litres/capita day.
Wastage can be assumed to be 30 per cent of production.
Therefore - total daily production needs to be: $1000 \times 100 \times \frac{100}{(100 - 30)}$ = 143 000 litres/day

= 143 m^3/day

The rate of filtration is 2.4m^3/m^2d (=0.1m/h)

Therefore - the filter tank needs to have an area of: $\frac{143}{2.4}$ = 56m^2 (7m x 8m)

Two parallel filters are required, each 7m x 4m, with a common divide wall.

Overall filter tank height is made up of:		
	above level of water	- 0.3
	water	- 1.0
	sand	- 1.2
	gravel	- 0.3
	underdrainage	- 0.15
	foundations	- 0.15
		3.10 metres

The Final Design

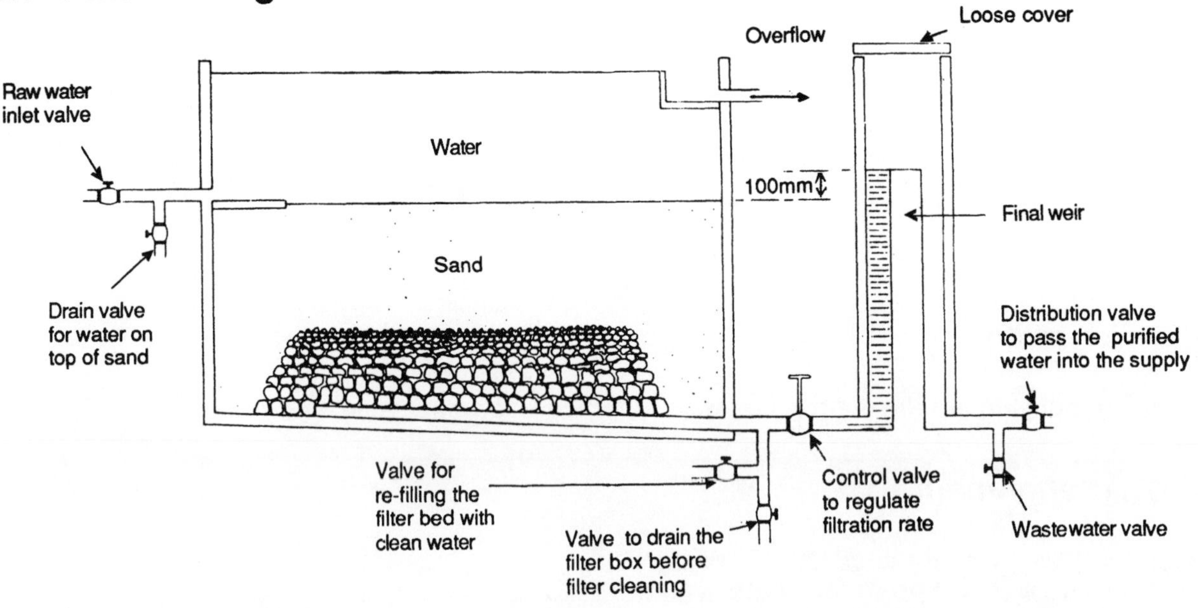

Ken Ellis, WEDC, Loughborough University of Technology, UK.
Claire Purvis, WEDC, Loughborough University of Technology, UK.

16. Sewerage

Technical Brief No. 10 dealt with WASTE STABILIZATION PONDS. Whether treatment is in ponds or in sewage plants (with percolating filters for activated sludge) sewage has to be carried in sewerage — a system of sewers.

SEWERS are normally circular pipes,

although some sewers have other shapes.

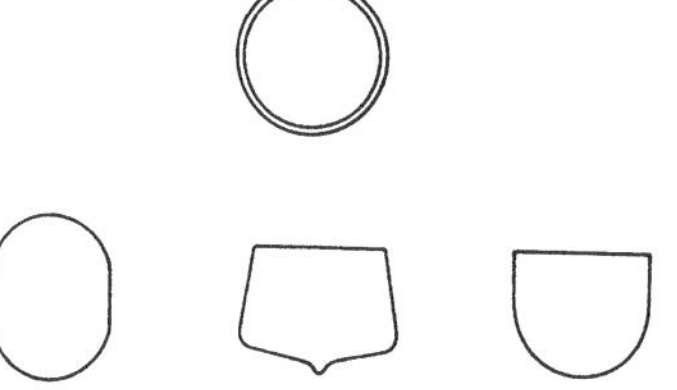

SEWERS are made from the following materials:

plastics	- usually small diameters (62-300mm) - care needed with storage
asbestos cement	- liable to damage during transport
clayware	- should be glazed/vitrified - can be made locally
bricks	- local kiln-dried bricks can be used
plain concrete	- requires smooth inside surface
reinforced concrete	- for large diameters (300mm or more)
cast iron	- for pumping mains and sewers under buildings

Foul sewage

Foul sewage consists of flow from water closets;

and wastewater from washing or bathing;

from washing clothes;

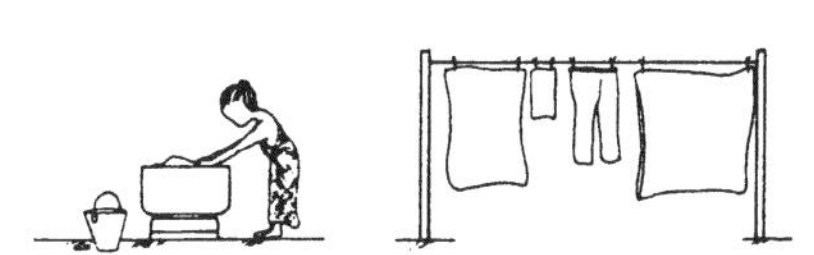

from preparing, cooking and serving food;

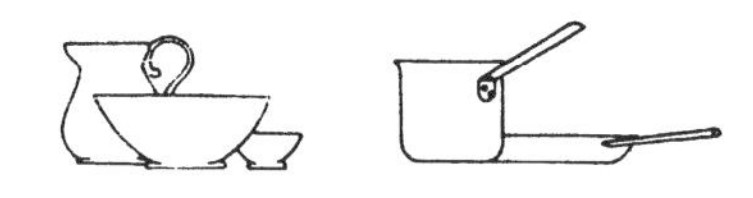

and other domestic, commercial and industrial wastewater.

SEWERAGE

The quantity of foul sewage depends on the amount of water supplied. If 75 - 200 litres per person per day is provided, the sewage flow may be taken as 80% of the water supply.

Greater water use is often due to garden watering, and the water used for this does not go into the sewers.

Most sewers are designed to carry foul sewage only.

Rainfall is removed in a separate system of storm drains or monsoon drains.

Nevertheless, some rain always gets into foul sewers.

If the groundwater table is above the sewer (either throughout the year or seasonally) some infiltrates into the sewer.

Sufficient capacity has to be provided in foul sewers to carry rainwater and infiltration water. Sometimes it is assumed that the maximum sewage flow will be some multiplier of the dry weather flow (dwf). 6 x dwf is often used for design.

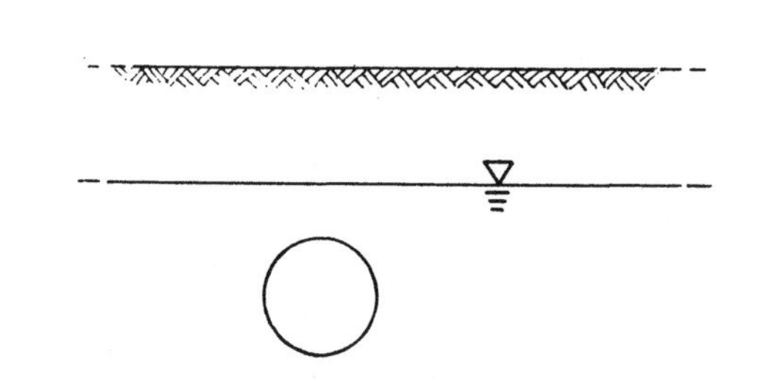

Gravity sewers

Normally sewage flows along sewers because they are laid with a slope (or 'gradient'). Where the ground has a natural slope, the depth of the sewers below ground level is often made constant. The sewers slope downhill towards ponds or treatment works in low-lying areas.

Where land is flat, sewers get deeper to maintain a downward slope. Deep sewers are expensive and difficult to construct so it is necessary to raise the sewage by pumping.

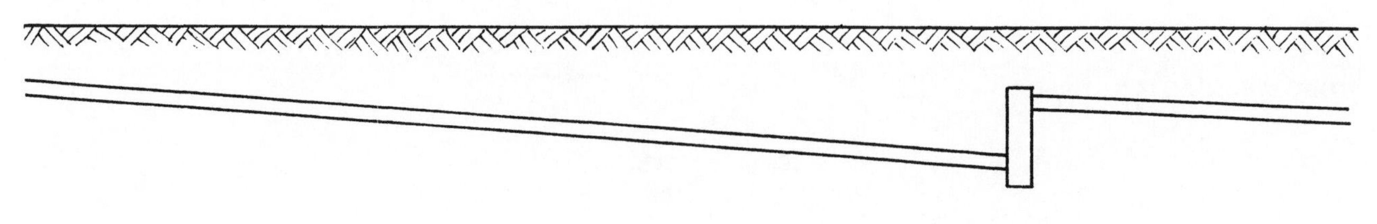

Pumping station

Pumping of sewage

* adds to the capital cost;
* introduces running costs for power;
* is liable to failure because of plant breakdown, shortage of fuel or electricity failure.

Design

1. The gradient should be sufficient to ensure that the velocity is at least 0.6 metres per second when carrying the maximum daily flow. Solids are then carried along the sewer. Minimum gradients to give this velocity are:

	Sewer Diameter			
Material	100mm	150mm	225mm	300mm
Plastic	1/185	1/305	1/500	1/710
Asbestos cement	1/180	1/295	1/485	1/685
Clayware	1/175	1/285	1/465	1/660
Smooth concrete	1/155	1/255	1/420	1/600
Rough concrete	1/85	1/150	1/250	1/370

2. The sewer capacity must be enough for the maximum expected flow (or 'design flow') of all the sewage entering upstream, allowing for rainfall and infiltration. Published tables or graphs are used to find the size of sewer needed to carry the design flow, taking account of the pipe material and the gradient.

3. The maximum daily flow of the foul sewage (i.e, excluding rainfall and infiltration) may be based on the maximum water supply.

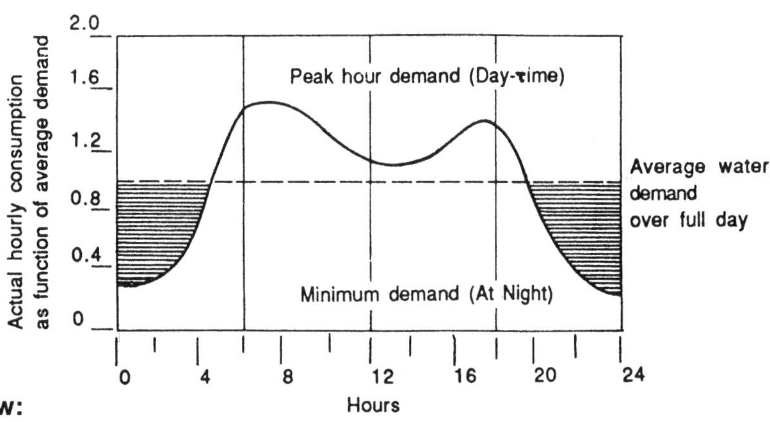

Ratio of maximum flow to average flow:

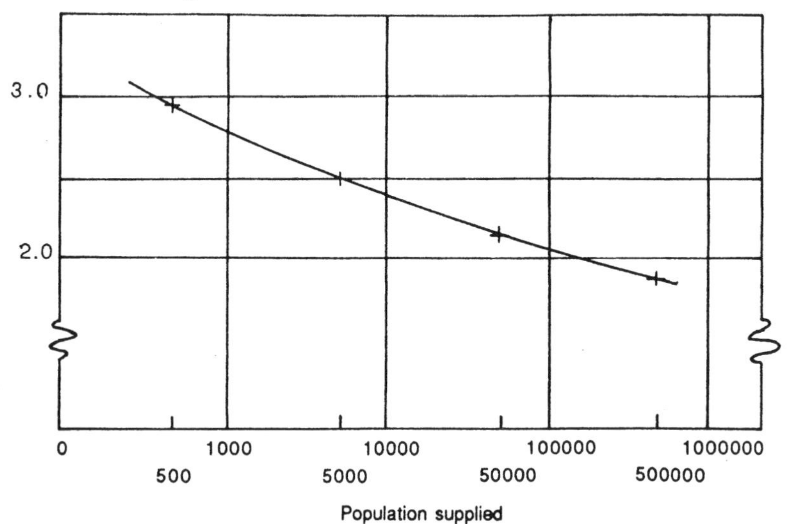

4. For a single house the maximum flow may be based on the simultaneous discharge of water fittings, for example, w.c. flushing while baths and sinks are emptied.
As more people contribute to a sewer there is less likelihood of them all using the system simultaneously and the ratio of maximum flow to average flow decreases.

SEWERAGE

Disadvantages and limitations of Sewerage

HIGH COST

Sewerage is the most expensive of all sanitation systems. Sewers alone may cost up to $1300 per person.

WATER USE

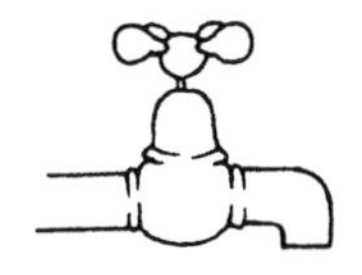

Good piped water distribution is essential. Sewers block if supply is intermittent or if the water supply fails during dry weather. Normally there must be a supply sufficient for 75 litres per person per day.

HOUSE CONNECTIONS

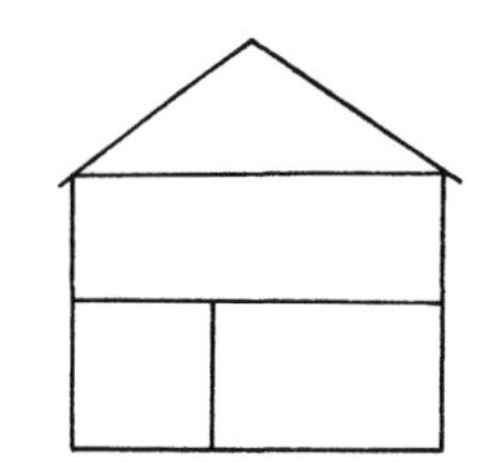

Every house must have piped water with multiple outlets to w.c., bathroom and kitchen. Internal plumbing is expensive.

GOOD DESIGN

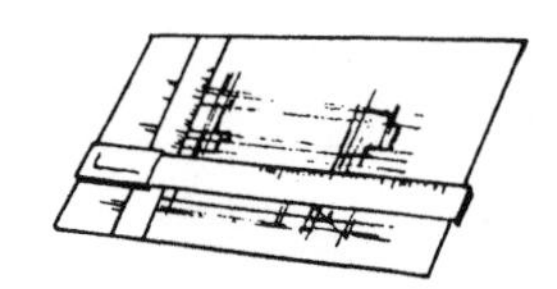

Sewerage systems must be designed by qualified technologists.

N.B. Engineers with conventional training based on industrialized country technology are often more capable of designing sewerage systems than low-cost sanitation options.

CONSTRUCTION

To ensure good operation and long life, sewer laying should be supervised by professional engineers or surveyors.

Unconventional Sewerage

Two systems with reduced diameter and reduced gradient have been installed in a few places.

1. SMALL BORE SEWERS carry effluent from tanks (septic tanks, aqua-privies or interceptor chambers). Solids are retained in the tanks, so there is less risk of sewer blockage, providing solids are regularly removed from the tanks.

2. **In high-density housing areas, CONDOMINIAL SEWERAGE uses conventional sewer pipes laid at a shallow depth within plots, normally behind the houses. Householders are responsible for clearing any blockage within their plots.**

John Pickford, WEDC ,Loughborough University of Technology, Loughborough ,Leics, LE11 3TU, U K.
Claire Purvis, WEDC, Loughborough University of Technology

17. Health, water and sanitation (1)

Safe drinking-water and sanitation are essential for the health and the development – social and economic – of families, communities and nations. Without water, humans die within a few days from dehydration. Death tends to be higher and life expectancies lower in areas with poor water and sanitation. In half of the developing countries, the Infant Mortality Rate is greater than 100/1000 live births and the life expectancy at birth is less than 50 years (WHO, 1986). About 40,000 children die everyday of diseases related to water and sanitation. The great disparity in the provision of essential facilities between the developed and the developing countries and, especially in the latter, between urban and rural populations is illustrated in Table 1.

Table 1: Water and Sanitation Urban/Rural Coverage

	Urban	Rural
% total population	30-35	65-70
% with adequate water supply	74	39
% with adequate sanitation	52	14

Source: WHO, (1986)

Poor water contains many micro-organisms

The need for clean water

Many people in the world, and especially those in developing countries, have water but still die because of its poor quality or the irregular quantity available to them. People, usually women and children, may spend many hours each day fetching and carrying water which is poor in quality simply because it is their only supply, and without it they die. Provision of clean water closer to homes may reduce the time and energy spent in water collection and increase the volume used. (See Table 2.) Yet this may not improve people's health if they continue living in otherwise insanitary conditions or have insanitary/unhygienic habits.

Poor environment

Table 2: Water Use

Source	Distance	Water usage L/Person/day
Well:	>2,500m	5
	<25m	15
Standpost:	<250m	15-35
Piped:	yard tap	75

This Technical Brief introduces health in relation to water and sanitation facilities and hygiene practices. Later Briefs will look in more detail at the effect of water and sanitation on health, disease (for example, malaria and schistosomiasis) and groups of infections (diarrhoeas and eye diseases).

HEALTH, WATER AND SANITATION (1)

Poor environment – poor health

Poor environmental conditions arising from unhygienic disposal of excreta and sullage and accumulation of solid wastes, contribute to the spread of disease. They lead to contamination of food and water supplies, either at source or in the home. They may encourage breeding of vermin and insects, further increasing the spread of disease.

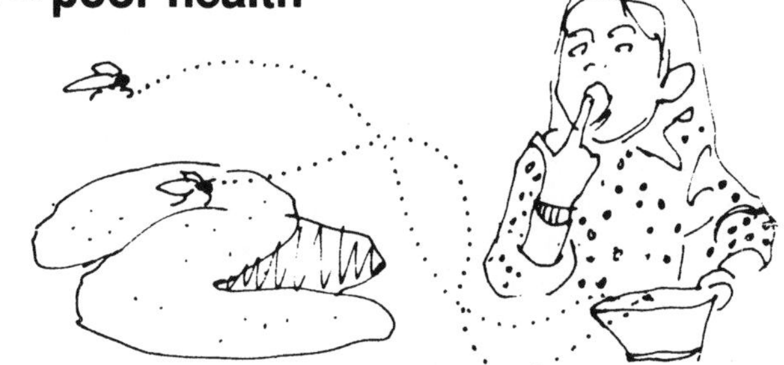

Faecal-oral transmission

Classification of water and excreta-related infections

The diseases caused by or related to water and excreta can be either non-infectious such as fluorosis, from high fluoride levels, or infectious such as cholera and malaria, which depend on disease-causing organisms (pathogens). In developing countries the infectious diseases predominate, hence they will be looked at in more detail.

As excreta is the major source of pathogens in water, most of the routes by which water and excreta-related diseases are spread (or transmitted) are the same. These routes can be used to classify the diseases as outlined below with examples and some control measures being given for each class. In addition to management of water and wastes, measures may include control of vectors (mosquitos for malaria or snails for schistosomiasis) or immunisation and other medical-based methods.

WATER-BORNE	pathogens are present in water supplies	**WATER-WASHED (WATER-SCARCE)**	spread of the pathogen is affected by amounts of water available for hygiene
example:	diarrhoeal infections, cholera, typhoid	example:	scabies, trachomas, pinworm infection
control:	water quality, hygiene education	control:	water quantity, soap hygiene education
WATER-BASED	the pathogen must spend part of its life cycle in aquatic intermediate host or hosts	**WATER-RELATED INSECT VECTOR**	the pathogen is spread by insects that feed or breed in water (flies and mosquitoes)
example:	1 guinea worm infection 2 schistosomiasis 3 lung fluke infection	example:	malaria, yellow fever, Bancroftian filariasis Onchocerciasis
control:	excreta disposal (2,3) water quality (1) water access (1,2)	control:	surplus water drainage and management, insecticides
SOIL-BASED	the excreted organism is spread through the soil		
example:	hookworm infection	control:	excreta disposal

Faecal-oral transmission
Although water-borne infections may be spread through water, the majority are spread by other routes, i.e. on hands, clothes, food, eating and drinking implements etc. As organisms from faeces reach the mouth by one of these routes the transmission is often called faecal-oral. Water-washed infections come into this category if the source of pathogens is excreta, e.g. pinworm infection. However some water-washed infections, such as scabies and trachoma, are not related to excreta but solely to personal hygiene and hence quantity of water rather than quality.

Health and hygiene

For improvements in water and sanitation to give maximum reduction in disease, some changes in hygiene practices are often necessary. Both personal and domestic hygiene practices are involved and include:

Water care – health care
Clean vessels should be used for collection and water should not be run over the hands into the vessel. Covering storage containers protects water from possible contamination by birds or animals. The drinking-water supply may also be kept separate from water for cooking and washing.

The area around the collection point should be kept free of wastes and standing water. If water does collect in pools or ponds for sufficiently long periods, insects, particularly mosquitoes, will be able to breed and increase the spread of the disease. If the ground around the collection point is moist, worm eggs and larvae can develop and infect other people collecting water.

Covered storage containers

Cleaner latrines – better health
Unless it is kept clean, even a well built latrine can increase the spread of disease. Hence the expenditure in time, labour, materials and finance, aimed at improving sanitation and health may be wasted. If the slab is soiled and water or urine is allowed to collect it can become a feeding and breeding site for flies and mosquitoes. Also eggs and larvae of intestinal worms deposited on the slab may be able to infect other users of the latrine.

If the latrine is structurally unsafe, unsightly or smelly and attracting insects, people will be discouraged from using it. They will revert to former practices, (e.g. open-field defaecation) and the health risks associated with them.

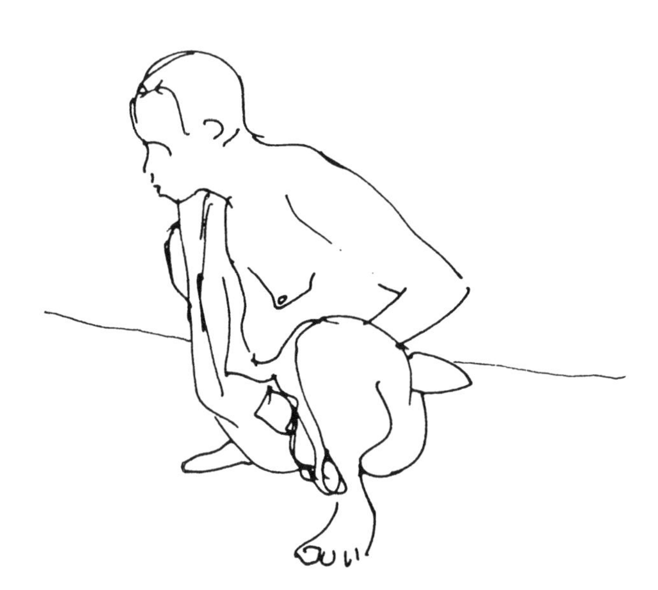

Open-field defaecation is a serious health risk

Education on the importance of both latrine hygiene and maintenance skills is therefore vital. One important part of such education should be that children's excreta can be very infective. It needs to be disposed of hygienically (e.g. in the latrine if the child is unable to use the latrine correctly or is too young to do so).

Disposing of children's excreta

Regular washing of clothes and skin

Failing to wash hands after attending to children's excreta or after using the latrine can lead to faecal contamination of food, water and clothing which in turn, may lead to illness. Good personal and domestic hygiene is more likely if water supplied or collected is greater than 35 litres per person per day and is located within 250 – 500 metres of the home.

The use of soap when washing hands and clothes is an important aid to good health. Soap can be made easily and cheaply from locally available materials (See Technical Brief No. 8). If not available, soap can be replaced by ash for washing.

Washing clothes

Planning water and sanitation projects

Knowledge of the diseases endemic in a particular area is an important factor in the selection of water sources and sanitation provision. Engineers should be aware of the locally common diseases and ensure that the facilities they are involved in providing do not increase rather than decrease the incidence of disease. For example, surplus water around wells and standposts should be drained, and disposal methods for sullage and waste water should be included in plans, and the customs and economic capabilities of those who will use the facilities should be taken into account.

For further information:

Cairncross, S. and Feacham, R. (1983) *Environmental Health Engineering in the Tropics, UK,* John Wiley & Sons.

Feachem, R.; McGarry, M.; Mara, D., (1977) *Water, Wastes and Health in Hot Climates, UK,* John Wiley & Sons.

T.A.L.C. (Teaching Aids at Low Cost), PO Box 49, St Albans, Herts AL1 4AX, UK.

Technical Brief No. 2 — *Introduction to Pit Latrines 7 — The Water Cycle.*

Technical Brief No. 8 — *Making Soap.*

W.H.O. (1986) The International Drinking Water Supply and Sanitation Decade. Review of National Progress (as at December 1983), CWS Unit, Environmental Health Division, W.H.O. 1211 Geneva 27, Switzerland.

Text: Margaret Ince Illustrations and design: Rod Shaw

WEDC, Loughborough University of Technology, Loughborough, Leicestershire LE11 3TU, UK.

18. Water testing

The objective of bacteriological testing of water is to detect and determine the concentration of faecal bacteria in water supplies in order to:

a) check that the supply is free from pathogenic (disease-causing) organisms and therefore safe to drink,

b) assess faecal pollution of supplies.

The number of organisms excreted into the environment each day by every living creature is enormous.

At present it is not possible to identify pathogens in drinking-water quickly. The pathogens may only be present occasionally in the water although pollution by faecal matter may be occurring all the time.

Normal bacteriological procedures to test for pathogens in the water determine if faecal pollution has occurred. If faecal bacteria are present, then pathogens may be there too. (See Diagram A.)

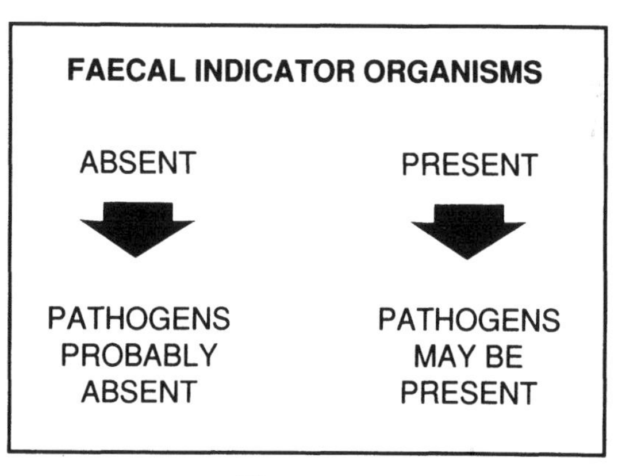

Diagram A

The most useful indicator organism for surveillance of bacteriological water quality are members of the faecal coliform group (See Diagram B).

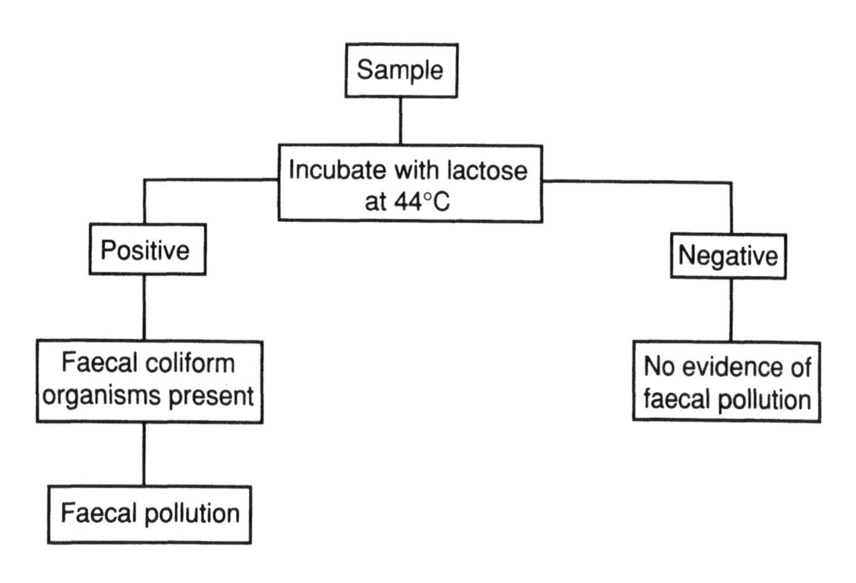

Diagram B

WATER TESTING

There are two main methods used to detect and measure indicator bacteria in water:

a) The membrane filtration method.
b) The most probable number (MPN) multiple-tube method.

a) Membrane filtration

The most useful method for testing faecal indicator bacteria in drinking-water is by membrane filtration. This procedure involves filtering a measured volume of sample (100ml), or an appropriate dilution of it, through a membrane filter which has a pore size of 0.45 µm. Micro-organisms are retained on the surface of the filter which is then placed on an absorbent pad soaked in a suitable selective growth medium (containing lactose) in a glass, plastic or metal petri dish. It is then incubated at 44°C for faecal coliform detection. Any bacteria able to grow will multiply to form visible colonies on the membrane filter surface. The number of colonies counted is expressed in terms of the number present per 100ml of original undiluted sample (See Diagram C).

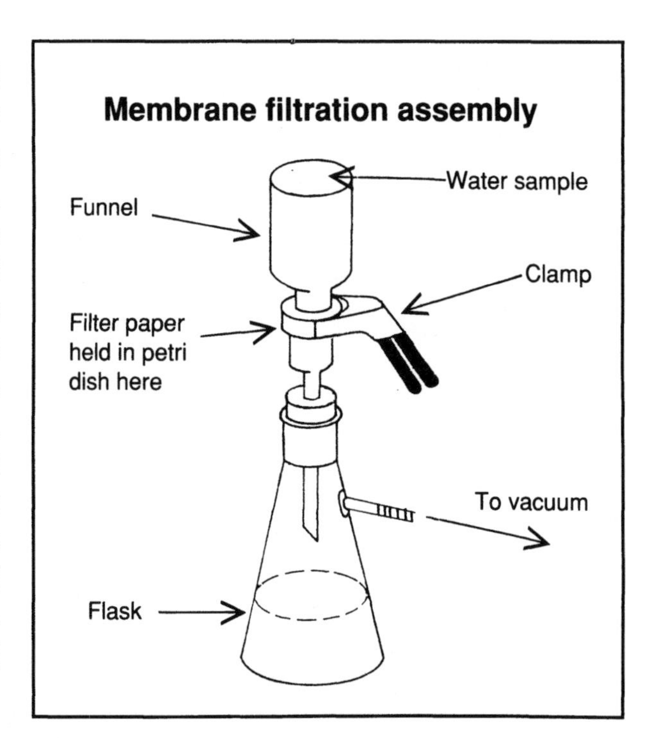

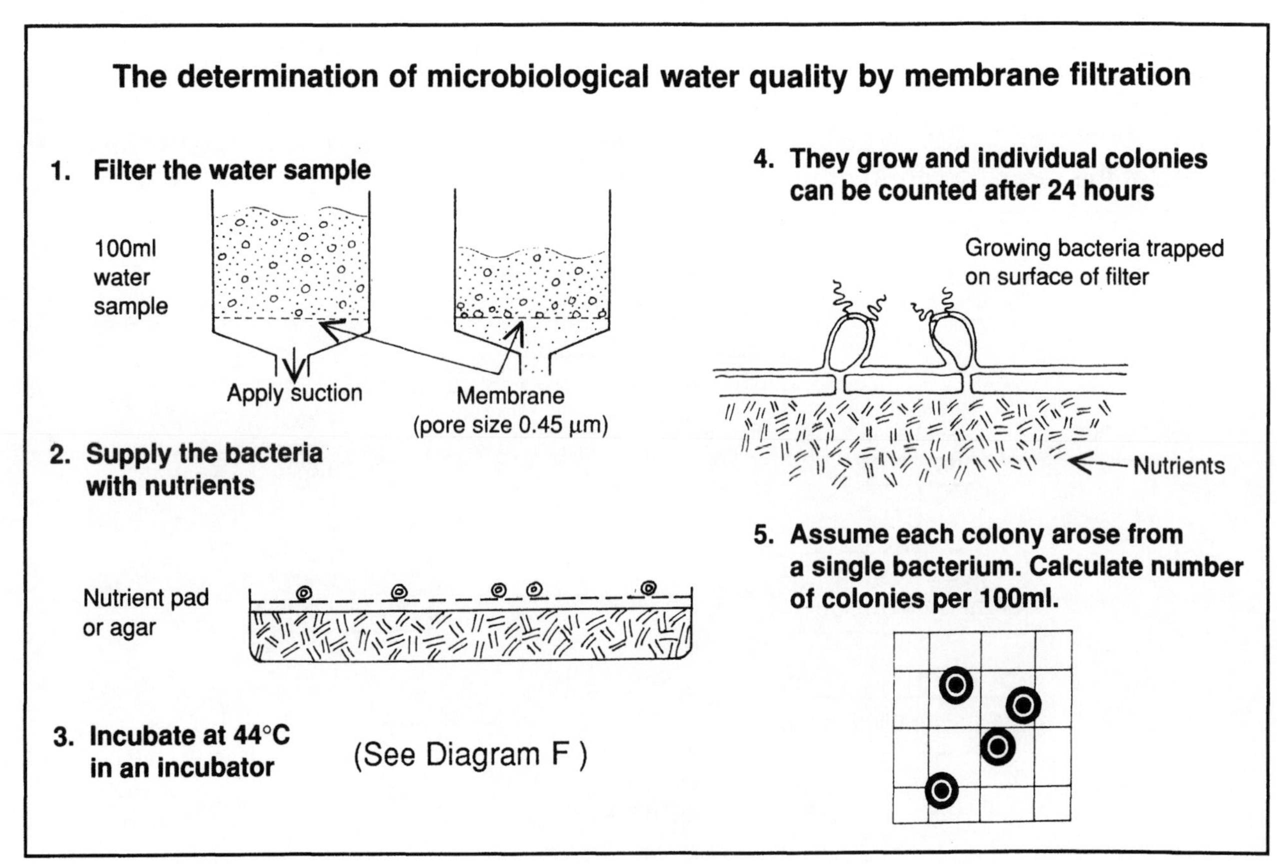

Diagram C

b) Most probable number (MPN) multiple-tube method

The multiple-tube method involves adding measured volumes of sample to sets of sterile tubes or bottles containing suitable liquid culture medium. Faecal coliform organisms produce acid and gas when incubated at 44°C. The gas production is detected by its appearance in a small 30mm inverted glass test tube (Durham tube) inserted into the tube or bottle before sterilisation (See Diagram D).

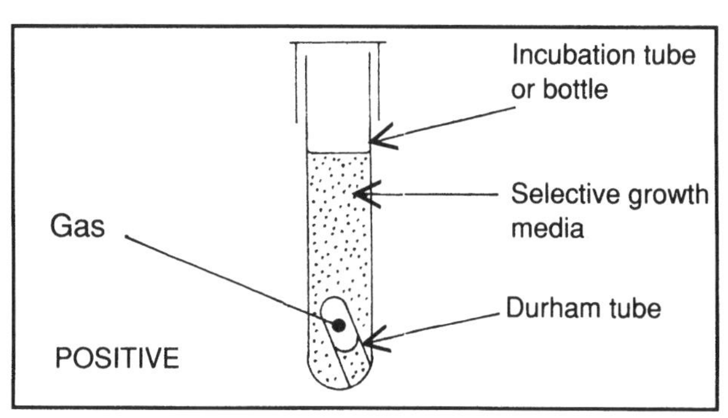

Diagram D

Acid is detected using various pH indicators. The number of tubes showing positive reactions is recorded at the end of the incubation period. An estimate of the most probable number (MPN) of organisms present in the original sample is obtained from statistical tables. A range of different dilutions should be used to ensure that both positive and negative reactions are obtained.

The multiple–tube fermentation or MPN technique is applicable to waters of all types and especially those with high turbidity. The equipment required is relatively cheap and simple, positive reactions are easy to read (See Diagram E). (One of the simplest procedures using 5 x 10ml of sample is described in detail by Mara in Cairncross and Feachem, 1978.)

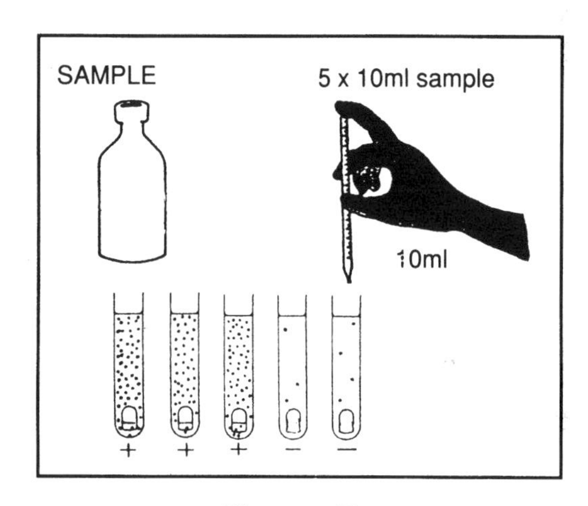

Diagram E

Incubators

The most difficult problem in operating bacteriological procedures in developing countries occurs with incubators. Battery-powered incubators have been developed in the past few years particularly for the membrane filtration technique (See Diagram F). They use an aluminium heating block, insulated and temperature controlled by electronic thermistors. The cheapest cost around £600 in Britain. Work is being carried out to provide cheaper heating systems for MPN and other procedures. For details, contact the author.

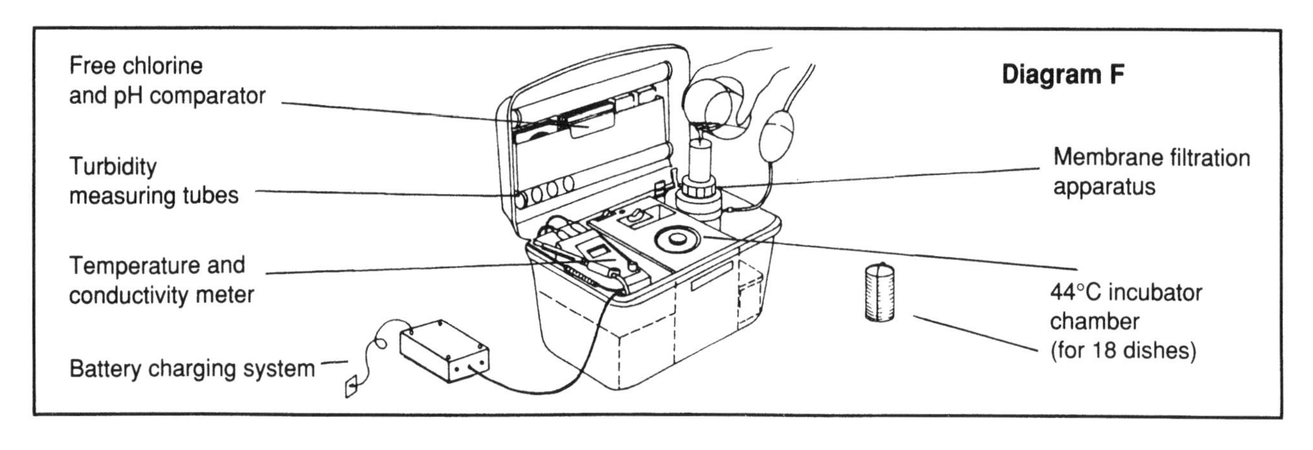

Diagram F

Autoclaves

Autoclaves are another essential part of a bacteriologist's equipment. In the absence of a commercial autoclave, ordinary domestic pressure cookers are suitable for sterilising sample bottles, screw-capped bottles, measuring cylinders, pipettes, petri dishes, tweezers, dilution and rinsing fluids and membrane filtration apparatus prior to use. They should also be used to destroy bacteria developed after the testing procedures have been carried out.

Membrane filtration apparatus

There are several systems available for filtering. These are made from polycarbonate plastic, glass or stainless steel. They come with syringes, pre-packed sterile selective growth media, petri dishes, absorbent pads, pipettes and membrane filters produced specifically for field use by several manufacturers such as Millipore, Sartorius, Oxoid, Nucleopore, Gelman and others.

General hygiene and cleanliness

The work area should be kept clean and free from dust and draughts. Hands should always be washed before and after sampling and analysing.

Samples

The sampling procedures are as important as the analysis. Samples for bacteriological analysis should only be collected in sterile bottles. Do not waste time and effort in testing samples which have not been collected in sterile bottles or improperly handled. Samples should be taken from a variety of points on a water distribution system. The reader is referred to more detailed texts on sampling procedures.

Water sampling is considered further in Technical Brief No.20.

References and further reading

APHA/AWWA/WPCF, *Standard Methods for the Examination of Water and Wastewater,* 16th ed, Washington, DC, APHA, 1990.

Mara, D.D. in Cairncross, S. and Feachem, R., *Small Water Supplies,* Ross Institute, London, 1978.

HMSO, *The Bacteriological Examination of Drinking Water Supplies,* HMSO, London, 1982.

Hutton, L.G., *Field Testing of Water in Developing Countries,* WRC Medmenham, UK, 1983.

Millipore Co. *Water, Microbiology, Laboratory and Field Procedures.* Millipore Corporation, Bedford, Mass. 01730, USA, 1984.

WHO, *Guidelines for Drinking Water Quality,* Vol. 3, WHO, Geneva, 1985.

Text: Len Hutton Illustrations and design: Rod Shaw
WEDC, Loughborough University of Technology, Loughborough, Leicestershire LE11 3TU, UK.

19. Health, water and sanitation (2)

Water and excreta-related disease

In *Health, water and sanitation 1* (Technical Brief No 17, 1988) diseases related to water and excreta were considered. The routes for the transmission of pathogens (disease-causing organisms) from excreta or blood of infected people to other, susceptible people were used to classify these diseases as:

1. Faecal-oral 2. Water-washed 3. Water-based 4. Insect vector 5. Soil-based

This Technical Brief will consider control of water and excreta-related diseases by interruption of transmission routes.

Measures to interrupt disease transmission

The main factors affecting transmission of water and excreta-related diseases are - water quality, quantity and management, excreta disposal, vector control and health education. Nutritional, medical and financial status are additional factors. Changes in these factors can control, reduce or even eradicate the incidence of disease by blocking transmission routes in one or more places. Some of the measures and their effects on disease transmission are considered below.

● Water quality

Water quality is particularly related to the faecal-oral transmission of disease.In addition, improved water quality alone can markedly reduce the incidence of Dracunculiasis (Guineaworm infection) - UNDP, 1988. For drinking-water quality guidelines similar to WHO's (WHO,1987), may be adopted. Maintaining quality involves care, not only in the selection and protection of sources and, where necessary, of treatment methods, but also in ways in which the water is collected at the source and then transported to and stored in the home.

Selection: Good boreholes provide reliable supplies of potable water. Groundwater quality is often better than that of surface water and therefore needs little or no treatment. Drilling boreholes, however, requires expertise in locating sites and operating rigs. The cost of operating and maintaining a borehole programme may be offset by savings on water treatment costs.

Drilling a borehole

Filtering water through a fine mesh cloth

Treatment: Processes to improve water quality can be simple, for example filtering through a fine mesh cloth to remove Cyclops, the Guineaworm vector, or complex, for example a series of sedimentation, filtration and disinfection processes.

HEALTH, WATER AND SANITATION (2)

Protection: Covering wells protects them from pollution by pathogens. These may enter the source in surface run-off, especially during the rainy season, with animals or humans entering it, or on dirty containers used to collect water.

Constructing a covered well

Storage: Drinking-water should be stored in covered containers separate from non-potable supplies to prevent contamination. Water should be collected in clean containers used only for this purpose.

● Water quantity

Water quantity affects transmission of water-washed diseases such as scabies and also some faecal-oral diseases such as typhoid. It may be associated with changes in health awareness and behaviour. If it is increased sufficiently, agricultural practices and hence nutrition may improve. Water quantity is also determined by water management measures (below) that reduce loss through leakage, equipment failure and accidents resulting from poor design and/or operation and maintenance skills.

Regular supplies increase personal hygiene

Rainwater harvesting

Hygiene: Regular reliable water supplies near to the home can increase the use of water for personal and domestic hygiene. Good hygiene helps reduce skin and eye disease and promotes practices that block transmission of other types of disease.

Nutrition: Raising water quantity for agriculture improves health through changes in quantity, variety and quality of foods. Better nutrition leads to better health. Surplus food may be sold and earnings used to improve the quality and quantity of life.

Sources: Improving present sources of water and locating new sources close to homes also helps ensure provision throughout the year. Collection and storage of water in reservoirs or other rainwater harvesting systems regulates the supply.

● Water management

Water management measures affect most transmission paths. This is apparent after flooding where incidences of diarrhoeal disease and insect-vector disease such as malaria increase markedly. In addition, other infections associated with poor living conditions (measles, coughs and tuberculosis) may also develop if water management is poor.

Drainage: Provision of storm water drainage reduces flooding and amounts of standing water where insects and pathogens may breed. However, design and maintenance of channels, by cleaning out debris, removing plants and repairing faults, is essential to reduce breeding sites for pathogens and vectors (snails, insects etc) in both drainage and irrigation systems.

Maintenance of drainage

Mending a pump

Operation and maintenance: Consistent care of pumps and distribution systems helps to reduce losses and ensure a regular water supply. Intermittent supplies can cause not only water contamination but also customer dissatisfaction and reversion to old, poor quality but 'reliable' supplies.

Wastewater: Adequate treatment is necessary for all wastewater. Sullage disposal systems, often neglected, should be included in all water supply schemes. Casual disposal can lead to soil conditions favourable for insect and hookworm development. Use of soakpits (separate or with latrines), drainage channels or piped systems are preferred.

● Excreta disposal

Transmission of all diseases where the pathogen is excreted will be affected by the method of excreta disposal. This will include all faecal-oral and soil-based diseases.Latrine and personal hygiene practices are also important in blocking transmission of faecal-oral disease.

Design: Containing excreta so that water and soil are not contaminated and flies cannot carry pathogens from it to food is an effective way of blocking transmission of much disease. A variety of latrine designs are needed to allow for different site conditions and for the social and cultural preferences of users (see Technical Brief No 2). Maintenance of disposal systems (the slab, superstructure and, if incorporated, emptying equipment) is vital.

Location: Siting the latrines downhill from wells and at sufficient distances to prevent infiltration are critical, otherwise pathogens from excreta will contaminate the water source.

A mud and wattle latrine

HEALTH, WATER AND SANITATION (2)

Spraying for vector control

● Vector control

Control will depend on the vector (insect, mollusc etc.), but the purpose is to interrupt the life cycle of the vector. This will include measures to reduce breeding sites, i.e. all water and excreta management methods, and chemical and biological measures directed at specific stages of vector development. This is especially important for malaria and schistosomiasis control.

Health education = better health

● Education

Health awareness has to be learnt. Knowing transmission paths for locally endemic diseases increases appreciation of measures to block transmission. Health education can use all media methods available, oral, visual and aural, and different sites - schools, clinics and community meetings. Television and radio reach larger audiences and have been used in campaigns to increase use of oral rehydration salts (ORS) for treating diarrhoeas. Teaching health awareness promotes changes in behaviour and attitutudes that affect incidence of disease. Integration with preventive measures, such as water and sanitation provision and child immunization programmes, and with curative health (treatment of those already ill) leads to good health. This requires awareness at individual, community, national and international levels.

The environment and human health

Human health, therefore, depends on many factors in the environment, including climate; composition of air; water for drinking, cooking, bathing, agriculture etc; disposal of excreta and solid wastes; housing quality and density; availability of types of food and social and cultural practices. Disease, as well as good health (social, physical and mental) depends, therefore, on how these factors interact with each other and with people wherever they live.

For further information:

Cairncross, S. *Waterlines* ,Vol 7. No.1, 1988.

Peavy, H.S., Rowe, D.R. and Tchobanoglous, G., *Environmental Health Engineering.* McGraw-Hill International, 1986.

Paqui, H. *Fighting the dreaded guinea worm.* World Development (UNDP) Vol.1, No.3, 1988.

de Rooy, Carel et al, *Guinea Worm Control as a major contributor to self-sufficiency in rice production in Nigeria.* Lagos, UNICEF/Nigeria, 1987.

Technical Briefs
- No. 2, *Introduction to pit latrines*
- No. 7, *The water cycle*
- No. 17, *Health, water and sanitation I.*

WHO, *Guidelines for Drinking Water Quality,* WHO, Geneva, Vols. 1-3, 1985.

Text: Margaret Ince   Illustrations and design: Rod Shaw
WEDC, Loughborough University of Technology, Loughborough, Leicestershire, LE11 3TU, UK.

20. Water sampling

The objective of water sampling

This Technical Brief describes the sampling procedures for bacteriological testing of water (see *Technical Brief No. 18, Water testing,* for details of the analysis).

The objective of water sampling is to collect and deliver for analysis a sample of water representative of the bulk of water being examined. Sampling procedures are as important as the analysis. Care needs to be taken to ensure that there is no accidental contamination of the sample during sampling and transport.

Bacteriological analysis results are used to check the quality of treated drinking-water at taps and standposts on a distribution system. The other major use is to evaluate contamination by faecal bacteria of water sources, such as springs, boreholes, rivers and lakes.

Preparation and sterilization of sample bottles

1. Glass bottles of at least 200ml capacity with a ground-glass stopper or rubber-lined aluminium or plastic screw caps should be cleaned and washed thoroughly, then rinsed with distilled or de-ionized water.

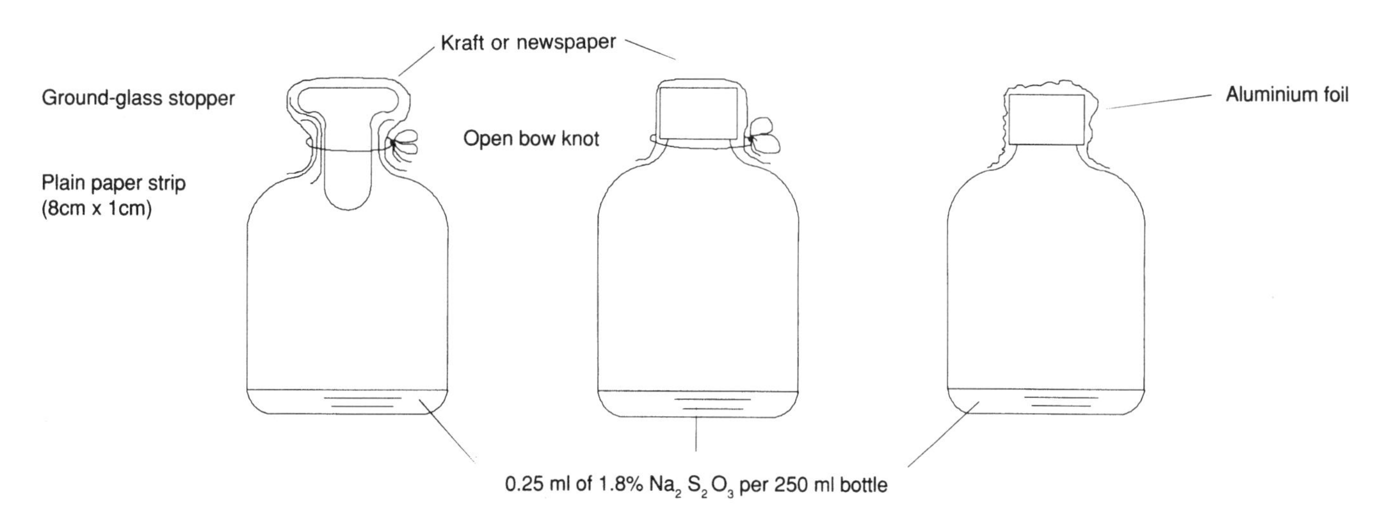

Sterile bottles and protected tops

2. Add 0.25ml of 1.8% solution of sodium thiosulphate ($NA_2S_2O_3 . 5H_2O$) per 250ml of bottle capacity to neutralize any residual chlorine.

3. Screw taps should only be loosely fastened prior to sterilization and then only tightened when cooled following sterilization. For ground-glass bottles only, a strip of paper should be placed in the neck of the bottle.

4. A piece of paper, preferably kraft paper or aluminium foil (although any thick paper, even newspaper, would do) should be fastened over the cap and neck of the bottle. This is tied with an open bow knot which can be released by a single pull.

5. The sample bottles are then sterilized in an autoclave at 122°C for 20 minutes or by heating in a dry oven at 170°C for one hour.

6. Allow bottles to cool and then tighten tops and store in a refrigerator until needed.

Sampling from a tap, standpost or pump outlet

A. Clean the tap
Remove any attachments from the tap that may cause splashing and, using a clean cloth, wipe the outlet in order to remove any dirt. If the tap leaks, it must be repaired before sampling. WASH YOUR HANDS.

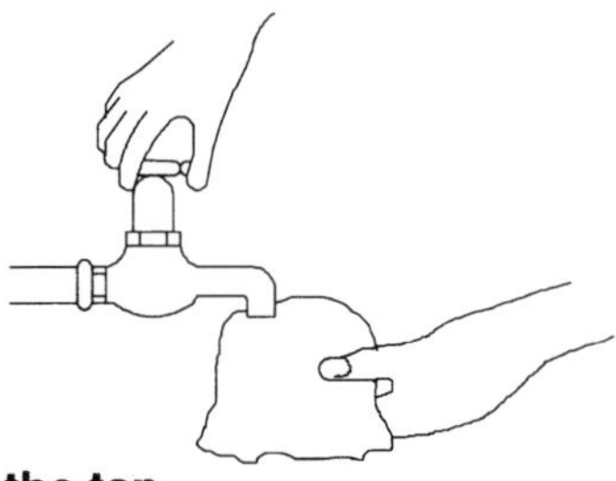

B. Open the tap
Turn on the tap at maximum flow rate and let the water flow for 1-5 minutes to clear the service line.

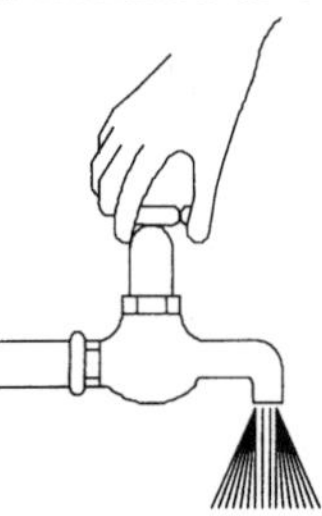

C. Sterilize the tap
Sterilize the tap for a minute with a flame from an ignited cotton wool swab soaked in alcohol; alternatively, a gas burner or cigarette-lighter may be used. If the tap is plastic, do not use a flame, a solution of hypochlorite will suffice. In some books tap sterilization is not considered important. However, if in doubt, sterilize the tap.

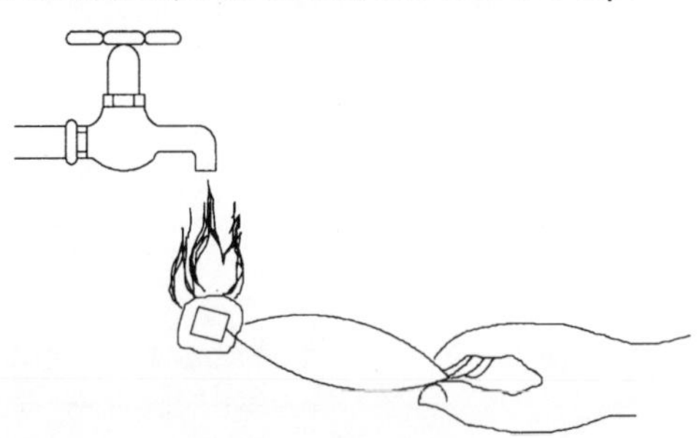

D. Open the tap prior to sampling
Carefully turn on the tap and allow the water to flow for 1-2 minutes at normal flow rate.

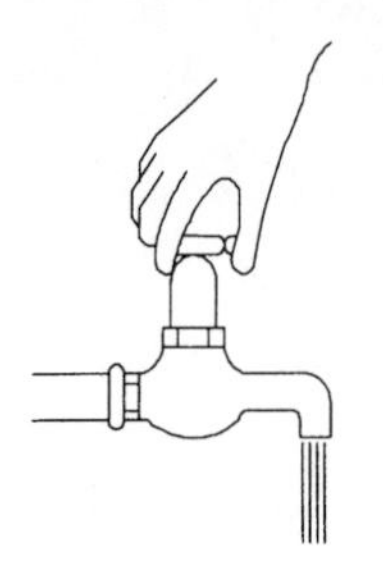

E. Open a sterilized bottle
Untie the string fixing the protective paper cover and pull out or unscrew the stopper, keeping your fingers on the paper.

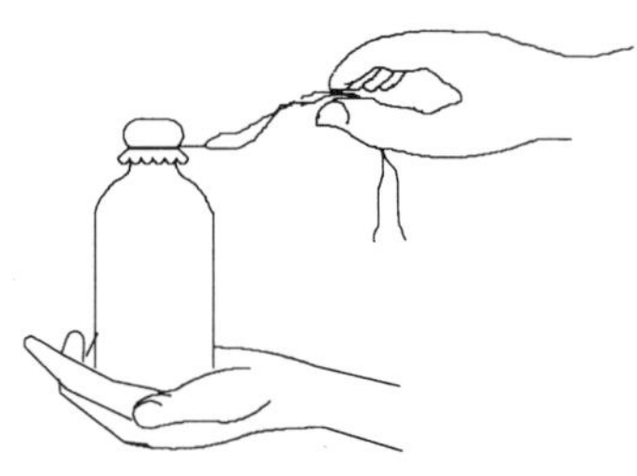

F. Fill the bottle
While holding the cap and protective cover face downwards (so as to prevent entry of dust that might carry micro-organisms), immediately hold the bottle under the water jet and fill to the shoulder of the bottle only.

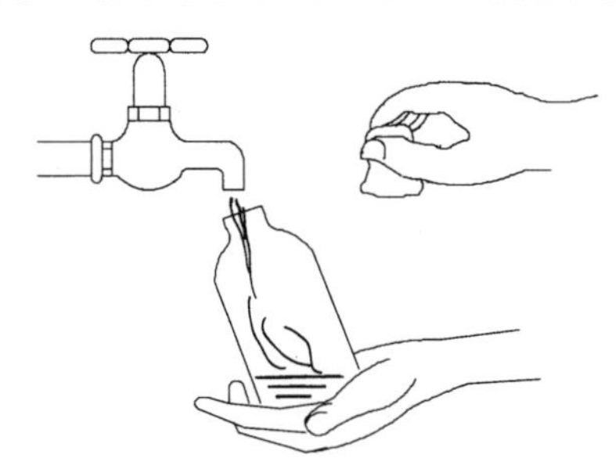

G. Stopper or cap the bottle
Place the stopper in the bottle or screw on the cap and fix the paper or foil covered cap in place with the string. Turn off the tap. Thank the consumer for the use of the tap.

Place a label on the bottle. Label with location (house no., road etc.), time of sampling, date and sampler's name. Place it in a transport box and return it to the laboratory within 24 hours if stored in melting ice.

Sampling from a watercourse or reservoir

Holding the bottle by the base, submerge it to a depth of about 30cm with the mouth facing slightly upwards. If there is a current, the bottle mouth should face towards the current. If there is no current, scoop the bottle away from your body. If the bottle is completely full, discard some water to provide some air space. The bottle should then be stoppered or capped.

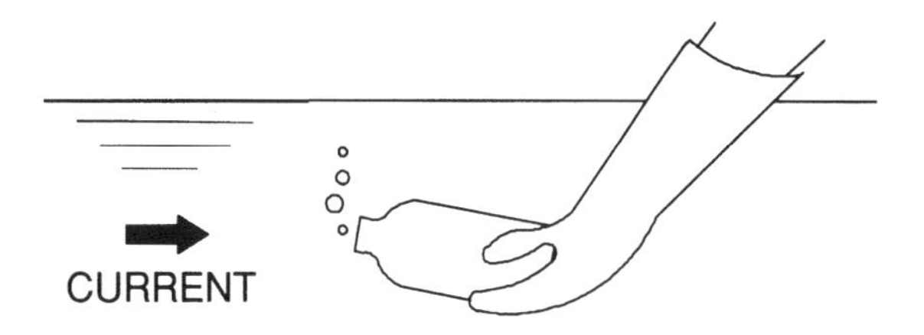

Sampling from wells, boreholes and water containers

For wells or boreholes equipped with a pump, the pump should be operated to clear any standing water in a water column (this pumping could be at least 20 - 30 minutes depending on the depth and diameter of the borehole) before the outlet pipe is sterilized using either a gas torch or alcohol flaming. Operate the pump for a further two minutes and take a sample in the flowing stream of water. For shallow open wells a weighted bottle or shallow sampling device may be used.

A. Prepare the bottle
With a piece of string, attach a stone of suitable size to the sampling bottle.

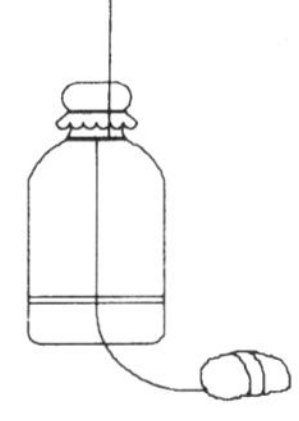

B. Attach bottle to string
Take a 20-metre length of clean string rolled around a stick and tie onto the bottle string. Open the bottle.

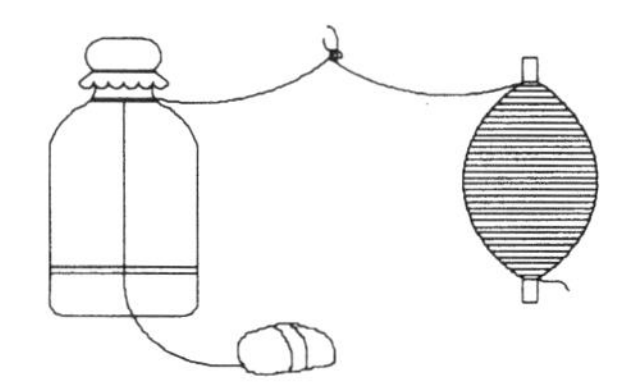

C. Lower the bottle
Lower the bottle, weighted down by the stone, into the well unwinding the string slowly. Do not allow the bottle to touch the side of the well.

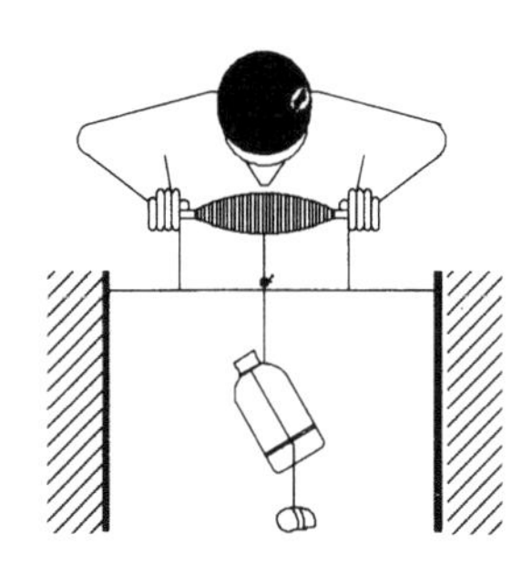

D. Fill the bottle
Immerse the bottle completely in the water and lower it to the bottom of the well.

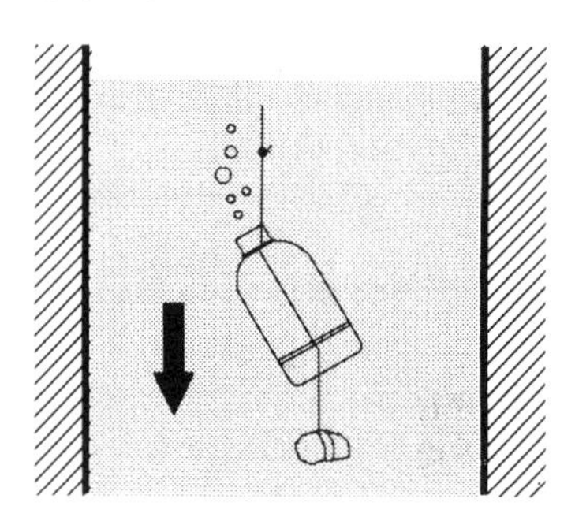

E. Raise the bottle
Once the bottle is filled, rewind the string round the stick to bring up the bottle. If the bottle is completely full, discard some water to provide some air space. Stopper the bottle.

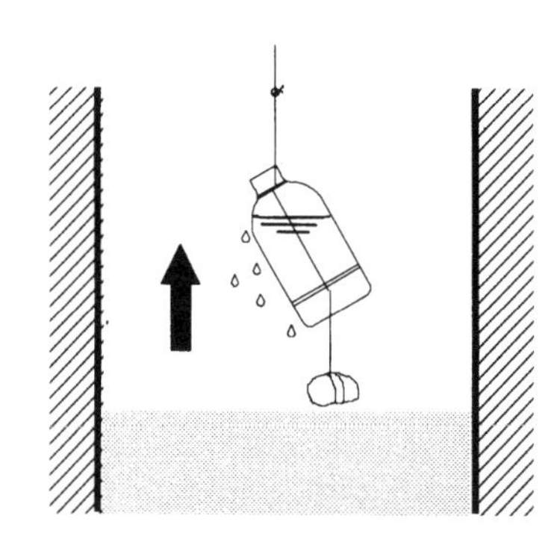

A sample from the water carrier's pot or bucket may be more representative of what is actually being drunk so take a sample poured into the bottle from that as well. If the water is stored in the house then the household container would also need to be sampled. Positive results from the house only would indicate poor hygiene in the home rather than polluted groundwater.

WATER SAMPLING

General points

DO:

- Collect the bacteriological sample first at sampling point.
- Only collect in sterile bottles.
- Keep the bottle closed until the sample is ready to be collected.
- Hold the bottle around the base.
- Carry some spare sterile bottles.
- Wipe the outside of the bottle.
- Resample if there is a possibility of contamination.
- Transport the sample in a cooled (0 - 4°C melting ice) covered container within 24 hours.
- Label the bottle with a waterproof marker pen with location, time, date and sampler's name.
- Test for chlorine residual on site by using a DPD or starch-iodide method. (Hutton, 1983.)

DO NOT:

- Contaminate the sampling point.
- Allow the top or neck of the bottle to touch anything.
- Collect samples in dirty bottles.
- Rinse or completely fill the bottle.
- Put yourself at risk from bilharzia: wear waterproof gauntlets or waders.

Frequency of sampling

Ideal frequencies related to populations are described in WHO's *Guidelines for Drinking Water Quality* (1985) but few countries can afford or are able to meet these recommendations due to a lack of trained staff and resources. The general criteria should be, however:

1. **To test as often as possible.**
2. **To test at as many points in the water network as practically feasible.**
3. **To keep the testing facilities fully employed until an acceptable frequency of sampling is obtained.**

There should be a concentration of sampling at points where maximum benefit will be obtained. It is essential that treated water entering the distribution system is monitored daily. Other points to be sampled are shown below. The minimum WHO recommendation on frequency of sampling for piped supplies is 1 sample per month per 5,000 people served.

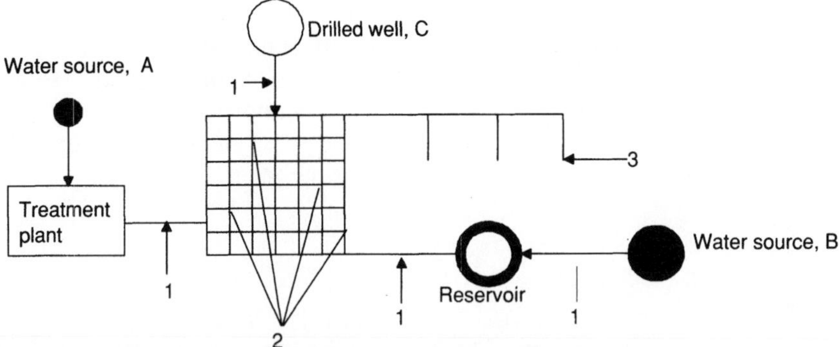

Typical sampling points. (Priority, 1-3)

For further information:

Hutton, L.G., *Field Testing of Water in Developing Countries,* WRC Medmenham UK, 1983.
Hutton, L.G., Technical Brief No. 18.
WHO, *Guidelines for Drinking Water Quality,* Vol 1 - 3, WHO, Geneva, 1985.

Text: Len Hutton   Illustrations and design: Rod Shaw
WEDC, Loughborough University of Technology, Loughborough, Leicestershire LE11 3TU, UK.

21. Slow sand filters (2)

Operation and maintenance

The effectiveness of slow sand filters depends very much on the style of operation and maintenance. A major advantage of this process is the limited number of tasks which must be performed, **but these must be carried out correctly.** (For design notes, refer to *Technical Brief No. 15.*)

Daily tasks

1. Ensure the depth of water in the reservoir above the sand is near the maximum. The level must be just at or very slightly below the overflow. **This level should not be allowed to fall.**

2. Adjust the treatment rate and the design flow by slightly opening the control valve (or slightly closing it if it was previously incorrectly set). The control valve is the outlet valve between the filter outlet and the final weir.

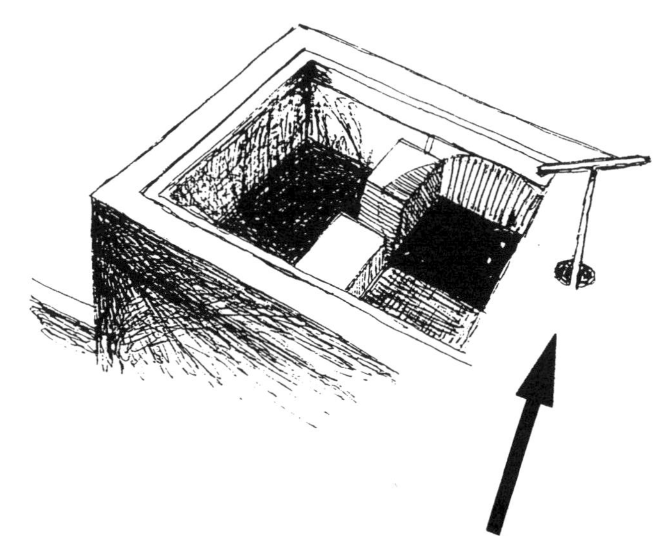

Adjust the outlet valve

The correct flow rate can be accurately judged by dipping the depth of water flowing over the final weir, or by checking the water depth over the V-notch weir set on the outlet weir.

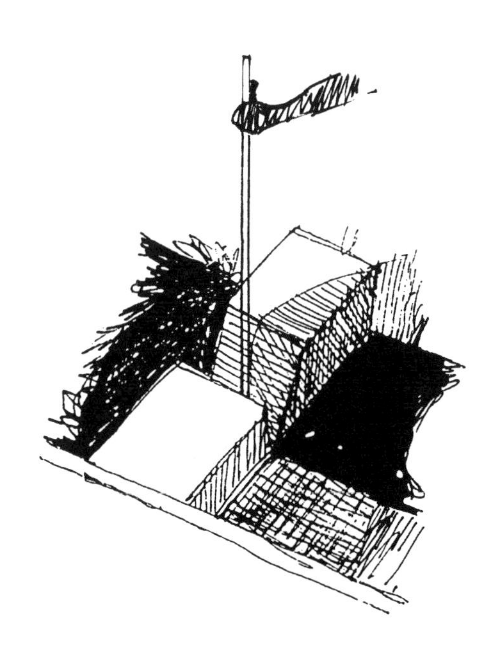

Dip the water flow

3. Observe the quality of the source water and of the filtered water. It is also helpful to check for any odour in both source and filtered water.

4. Note, in the records, the flow rate, condition of source and filtered waters and any unusual occurrences. Unusual occurrences might include unseasonal weather conditions, development of algae in the filter, rising *schmutzdecke* (filter-skin) or the illness of any of the operators.

Scraping the filter

Scraping becomes necessary when, with the maximum head of water available above the sand and the outlet valve fully open, it is not possible to obtain the design flow.

(When scraping the filter, it is essential to take measures to control the personal habits of the workers. There should be no spitting, urinating or defecating. Tools should be disinfected.)

1. Commence the scraping (or filter cleaning) operations by closing the inlet valve at the end of the day and allowing the filter to drain overnight.

2. Early next morning, run off any remaining water by opening the drain valve at the sand surface.

3. Continue to drain surface water through the filter until the level of water is about 100mm below the sand surface.

Swab down the filter walls

4. Swab down the walls of the filter box and remove any attacked algae. (See above.)

5. Begin scraping by using broad-bladed shovels or hoes to gather the *schmutzdecke* and any dirty sand into long ridges. The removal of a 25mm depth is usually sufficient. (See left.)

Gather the schmutzdecke *into ridges*

6. Remove the ridges of dirty sand and *schmutzdecke* from the filter. (See right.)

7. Smooth out the surface of the sand. Adjust the level of the sand surface drain and of the final weir if necessary and possible.

8. **Refill the filter from the bottom** until there is a depth of 150-200mm of water above the sand then stop filling from the bottom and resume filling in the normal manner.

Carry away dirty sand

Do's of slow sand filter operation

During filter cleaning:

- ***Do*** make sure that the cleaning operation is carried out quickly — one day is usually sufficient.

- ***Do*** make sure that there is a bath tray of clean water for all the personnel involved to walk through each time they enter the filter. Footwear should be provided.

Walk through a bath tray of water

Scare birds away from the filter

- ***Do*** control the personal habits of workers in the filter. No spitting, urinating or defecating.

- ***Do*** make sure that all birds are continually scared away from the exposed sand.

- ***Do*** refill the empty filter from the bottom.

Don'ts of slow sand filter operation:

- ***Don't*** dig up the whole of the sand bed during cleaning.
- ***Don't*** allow the level of the water in the reservoir to fall.
- ***Don't*** operate at varying rates.
- ***Don't*** allow people who are unwell to enter the empty filter during cleaning.
- ***Don't*** clean more than one filter at a time.
- ***Don't*** allow birds to foul exposed sand during cleaning.

General points

The following notes relate to *Technical Brief No.15, Slow Sand Filter Design,* and are intended to clarify design points.

1. The Gravel Layer

Instead of four layers of graded gravel illustrated, it is possible to use only a three-layer gravel system:

Top layer: 100mm depth of 1-1.5mm gravel
Middle layer: 100mm depth of 4-6mm gravel
Bottom layer: 100mm depth of 16-23mm gravel

2. The Uniformity Coefficient

The Uniformity Coefficient is the mesh size of a sieve in mm which retains 90% of the sand divided by the mesh size of a sieve in mm which retains 40% of the sand.

3. Inlet and outlet control

(i) The function of the inlet control system is to maintain a constant head of water above the sand.
(ii) The function of the outlet control system is to regulate the flow of water to the design rate.

4. Removal of viruses

A minimum depth of sand of 600mm is recommended to ensure the complete removal of viruses.
References: Poynter, S.F.B. and Slade, J.S., *The removal of viruses by slow sand filtration,* Prog. Water Technol., 9,75, 1977.
Windle-Taylor, E., *The removal of viruses by slow sand filtration*, Rep. Results Bact. Chem. Biol. Exam. Land. Waters, 44,52, 1969-70.

5. Eliminating wall effects

In order to eliminate wall effects it is necessary either to roughen the walls at the sand level during construction or ensure that the drainage system ends at least 600mm from the walls. It is preferable to take both precautions.

6. Inlet flow control

Inlet flow control by butterfly valve is advantageous if possible. It is not essential.

For further information:

Ellis, K.V., *Slow Sand Filtration.* CRC Critical Reviews in Environmental Control, 15, 4, 315-354, 1985.
Huisman, L. and Wood, W.E., *Slow Sand Filtration,* WHO, Geneva ,1974.
Slow Sand Filtration for Community Water Supply in Developing Countries: A Design and Construction Manual. IRC Technical Paper Series No.11, 1978.

Text: Ken Ellis Illustrations and design: Rod Shaw
WEDC, Loughborough University of Technology, Loughborough, Leicestershire LE11 3TU, UK.

22. Intakes from rivers

Introduction

A typical small water supply system requires less than about 200,000 litres per day, which is well within the capacity of small streams and alternative water sources, other than rivers. Indeed a river is not the ideal source of domestic water in many situations and an intake would normally be constructed only if there is no satisfactory alternative source such as groundwater (handpump), rainwater (catchment tank), or a spring (spring box). In contrast to these sources, water from rivers is liable to be polluted, and many rivers in the tropics and subtropics provide difficult conditions under which to construct an intake, for instance:

- They have a wide range of water levels between high and low flows, threatening to damage the intake at high flows, and leave it dry at low flows, and the intake has to operate satisfactorily over the whole range.
- They have a high sediment load ('silt'), especially at peak flows, which may block the intake.
- Scour and deposition can cause frequent changes to the bed and banks of the river channel, and may damage the intake or alternatively cut it off from the river.

Despite these problems, there are many circumstances where river water has to be used. The most suitable solution for village water supplies will often be a well or a series of wells along the river bank, provided that permeable materials of sand or gravel link the river and the well, without clay lenses to impede the flow. Water seeps to the well by sub-surface flow, and a distance of 50m from the river to the well should provide enough filtration to make the water safe to drink. Such wells also avoid problems of siltation and flood damage, and may still operate satisfactorily when the river is dry, by drawing on sub-surface water.

The wells can be drilled, jetted, augered or hand-dug and must extend some distance below the river bed level to give maximum year-round discharge. The top of the well should be above flood level, or sealed to prevent surface flood water entering the well and filling it with silt.

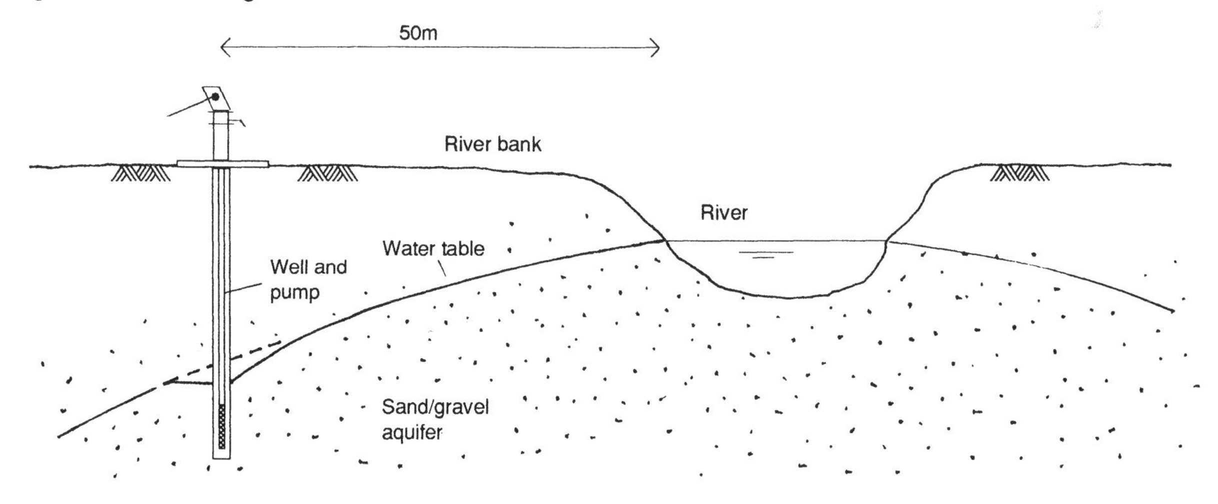

The remainder of this technical brief describes other types of intakes, for use when the above solutions are not suitable or would not deliver enough water. These intakes generally provide higher discharges, as would be required for large communities (or for small-scale irrigation schemes, where intakes are designed on similar principles). The illustrations show examples of the intakes, not standard designs.

Selection of an intake site

For all types of intake it is necessary to examine possible sites on the river and select a suitable protected, stable site on a stable, confined length of river, preferably upstream of a natural control section. Sites on the inside of bends should be avoided, to reduce sediment deposition and inflow, and side intakes should not be located where the river is wide and shallow, because of difficulties abstracting water at low flows without the expense of a long weir.

Exposed intakes

Pump with suction intake

On rivers with a stable bed and little variation between high and low water level, a pump may be set up on the bank, with its suction pipe down the bank ending in a screened intake below low water-level. Because of the suction limit of the pump, the difference in level between the mouth of the intake and the pump must be less than 3.5 to 4m.

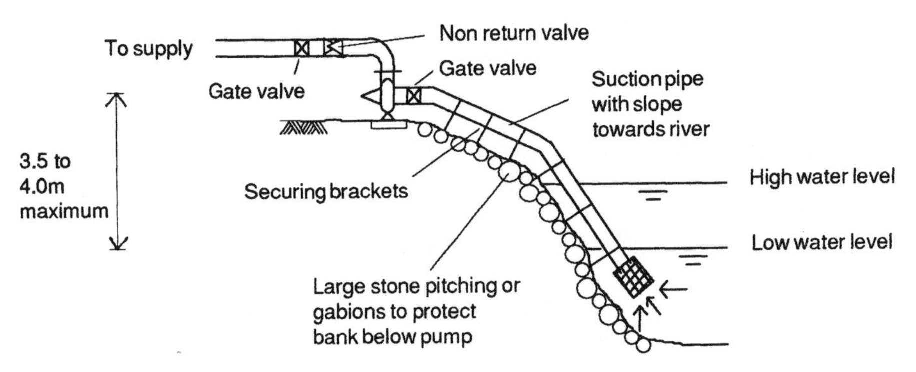

Intake with sump

For deeper intakes, a well-type pump may be set in a sump, with the inlet pipe through the river bank. In a simple intake, the pipe can be surrounded with a protective stone covering. In larger intakes a pier can be used to provide access to the intake, for instance to operate a valve or clean the screen. The pier also provides some protection, and could be fitted with a bar screen to exclude debris from the intake. A duplicate inlet pipe, sump and pump may be provided to make it easier to desilt the sump and maintain the pump.

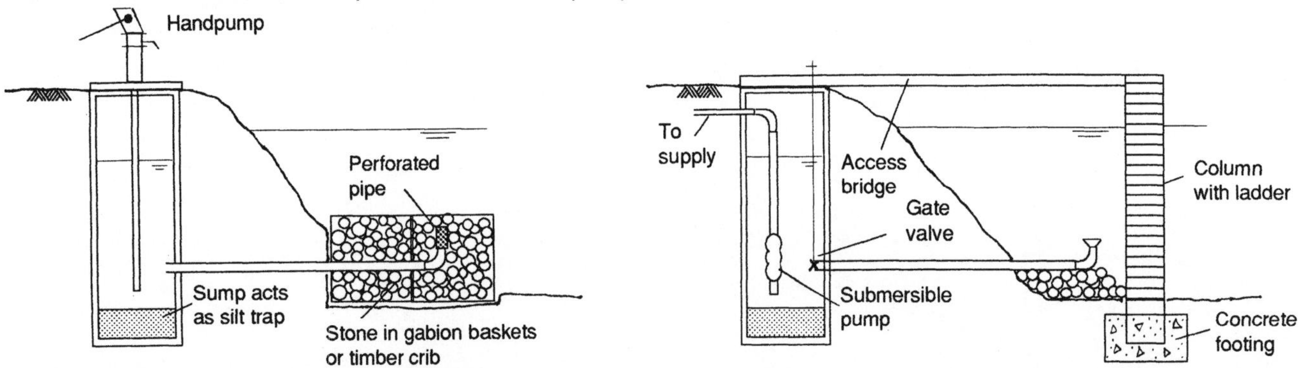

Problems can arise with all of the above intakes if the river bed is unstable (for example, gravel), or the river is very shallow at low flows. It can be difficult to set the intake level to be safe from silting up during floods, but still able to abstract water at low flows. A weir can assist with this (see opposite) or a floating intake can be used.

Floating intake

A floating intake has the advantage that it abstracts water from near the surface of the river, thereby avoiding the heavier silt loads carried on the river bed during floods. The danger with this type of intake is that children at play or floating debris such as tree trunks can damage the floats or cause the securing cables to break, making the intake inoperative. The example below shows a floating intake used as a temporary installation.

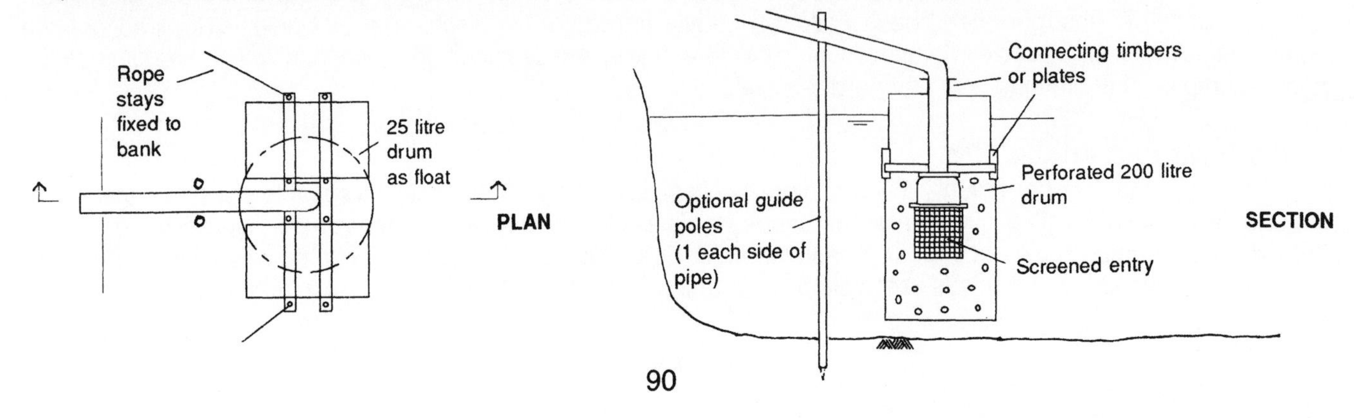

Protected side intake

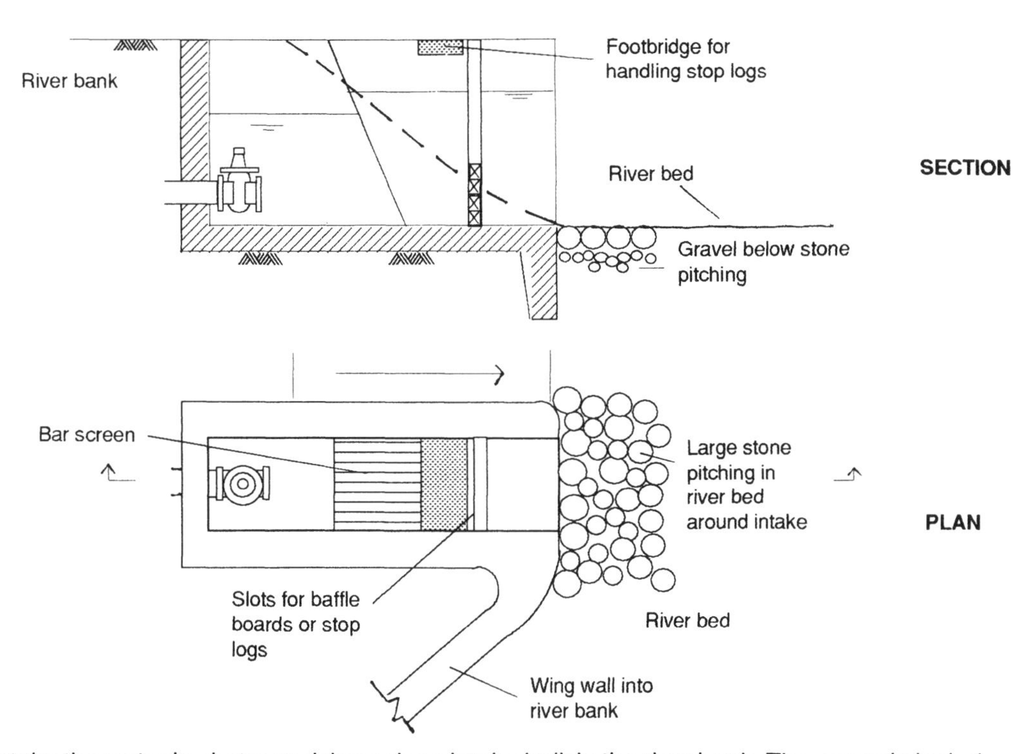

In the side intake the water is abstracted through an intake built in the river bank. The example includes wing walls into the bank and large stone pitching to protect the intake against floods and scour. A duplicate intake may be provided to facilitate maintenance.

Screens are used to prevent debris and large stones from entering the intake. A screen consists of a row of vertical steel bars, inclined at an angle of about 60 degrees to allow the screen to be cleaned by raking from above. A typical design uses bars of 25mm in diameter with a spacing of about 100mm, sized to give a velocity through the screen of about 0.5 to 0.7 metres per second.

The most important operation and maintenance tasks on this type of intake are:

- **To check the screens and rake them clear**
- **To clear any sediment which is deposited at the intake or its approach channel**

If the abstracted flow is a significant proportion of the dry season river flow, then it may be necessary to build a low weir across the river to divert the required flow to the intake.This could be a temporary boulder and brushwood weir, preferably combined with existing boulders on site. However, if the bed is unstable and liable to scour down, then a permanent low weir or sub-surface dam may be needed to maintain the water level above the intake. A possible design is shown below, suitable for rivers with a gravel bed.

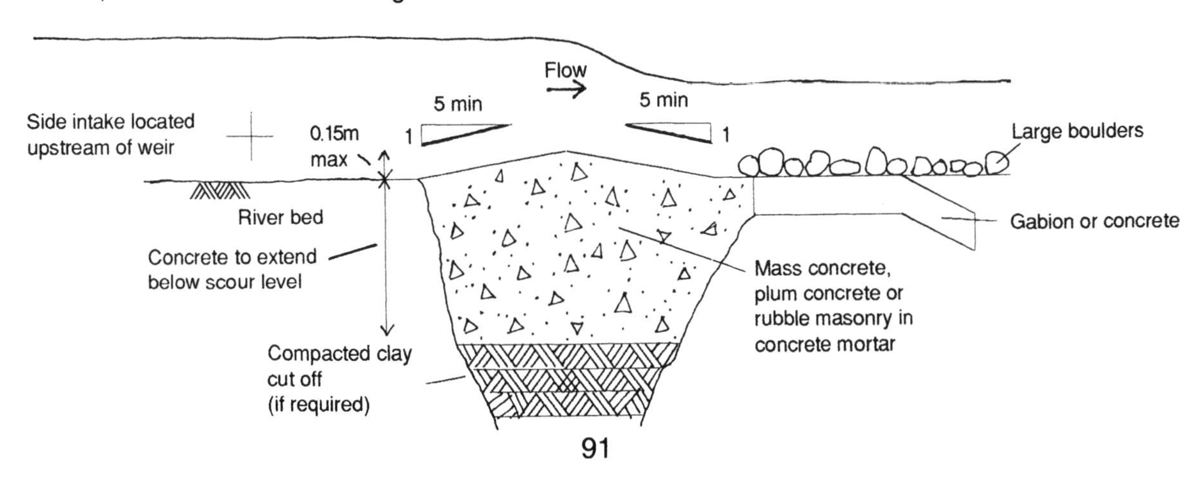

Infiltration galleries

The infiltration gallery draws on sub-surface flow, like the well on the river bank, and provides some filtration of the water. However, infiltration galleries are more difficult to design and construct than wells.

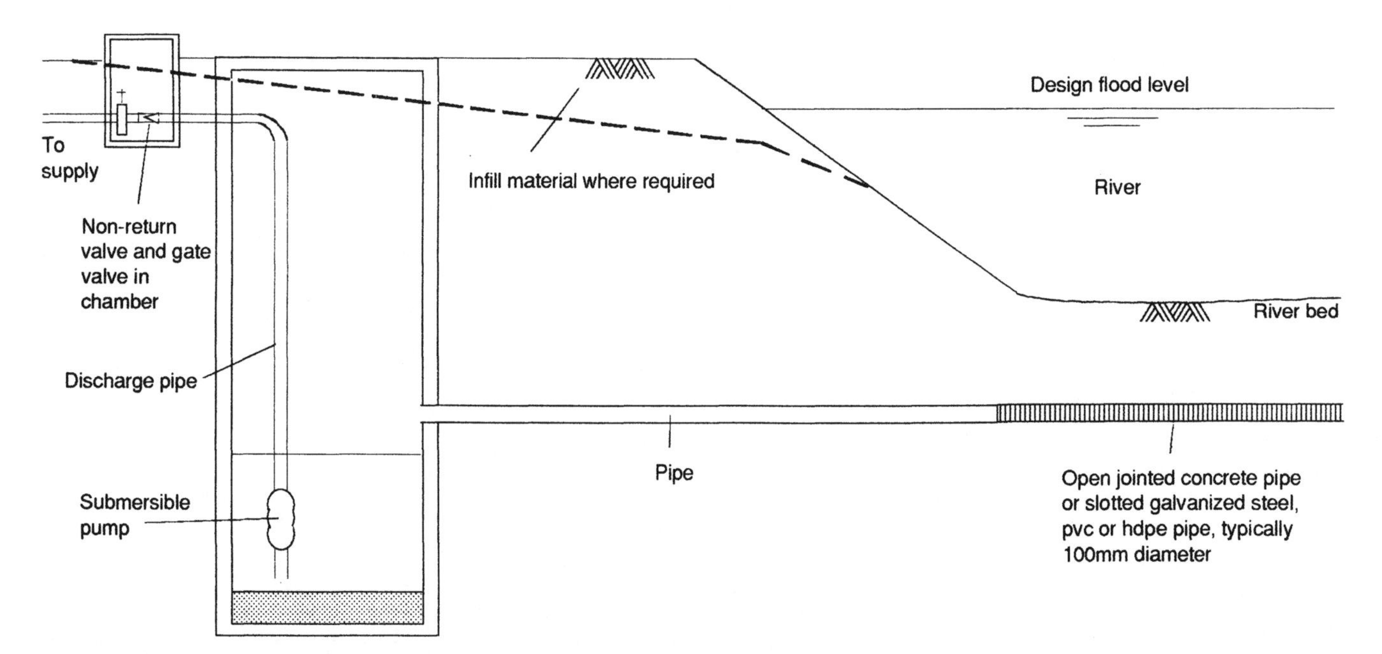

The infiltration gallery comprises an open jointed or slotted pipe laid below the river bed at a depth where it is safe from damage by scour. It is most suitable for use in river beds of medium to coarse sand, and in stable or degrading river sections where no sediment accumulation occurs. It is important to create and maintain an effective graded gravel filter around the slotted pipe, to prevent blockage of the filter or the pipe; for long life intakes, the filter should be designed in detail (see *Design of Small Dams*, p. 235). Blockage problems have led to the failure of some infiltration galleries, particularly in rivers carrying fine sediment. Construction involves excavation of a deep trench in the river bed which may be difficult and dangerous, and normally a de-watering pump is required. Alternatively, pipe jacking techniques can be used to drive the pipe horizontally through the bed, but without the safeguards of a filter. Infiltration galleries can also be constructed in the river bank, in a similar way.

A typical yield is said to be more than 15 litres per minute per metre length of gallery, but this depends on the difference in water level between river and sump. In dry river beds, this can be increased by constructing a sub-surface dam.

Another approach is to construct a sand storage dam on the river bed, incorporating an abstraction pipe and graded filter. As sediment is trapped behind the dam, the bed level rises and the pipe gradually comes to act as an infiltration gallery.

For further information:

Hofkes, E H (Ed), *Small Community Water Supplies,* IRC Technical Paper No 18, 1983.
Nilsson, Ake, *Groundwater Dams for Small-scale Water Supply,* IT Publications, 1988.
US Bureau of Reclamation, *Design of Small Dams, 2nd Edition,* 1979.
Technical Brief No.24, *Groundwater dams.*

Text: Ian Smout Illustrations and design: Rod Shaw
WEDC, Loughborough University of Technology, Loughborough, Leicestershire LE11 3TU, UK.

23. A guide to sanitation selection

Drawings 1-6 illustrate the principal options for on-site sanitation. The *Guide to Sanitation Selection* on the centre pages may be used to determine which option is likely to be most effective according to the method of anal cleansing, water availability and willingness to pay.

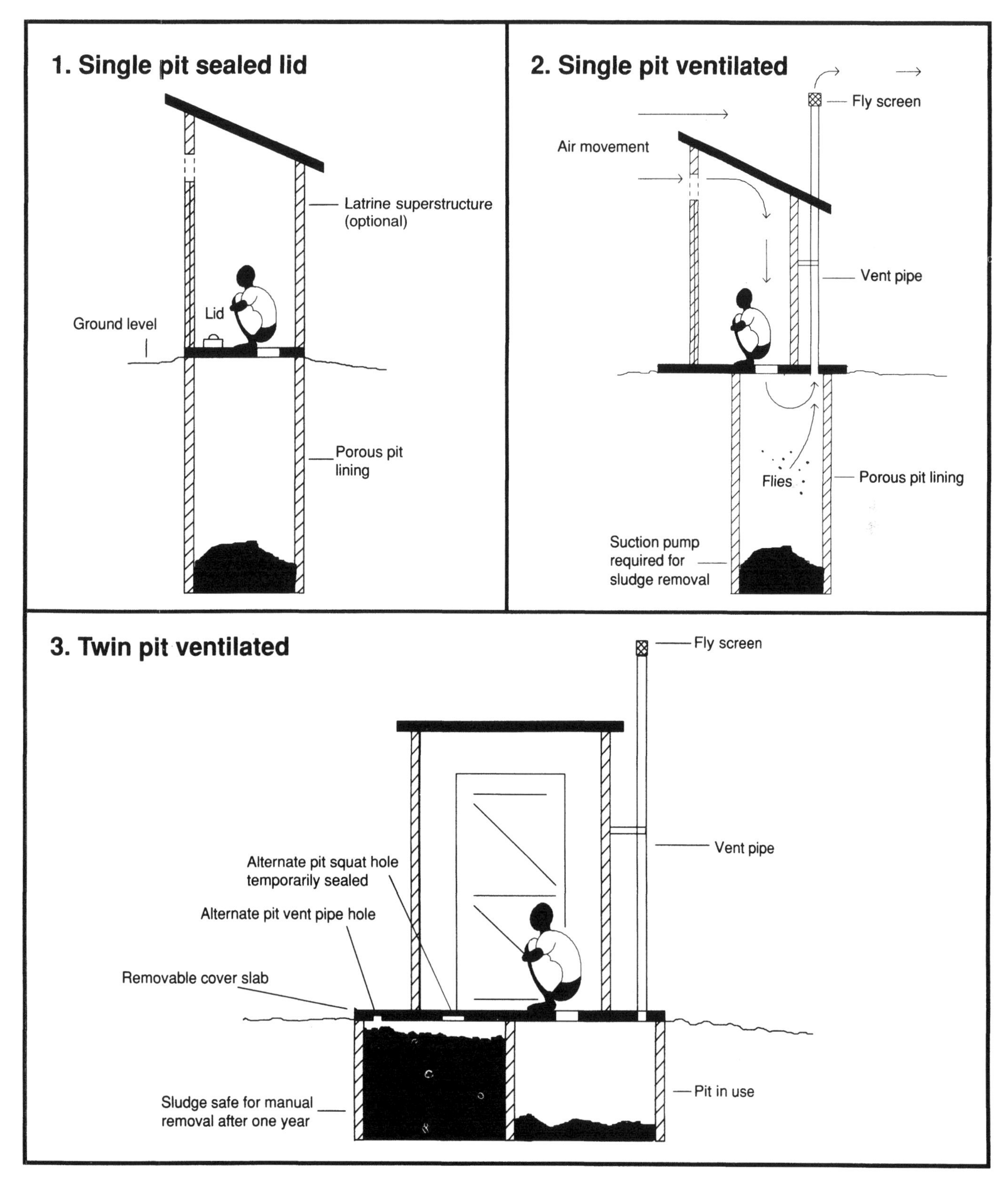

A GUIDE TO SANITATION SELECTION

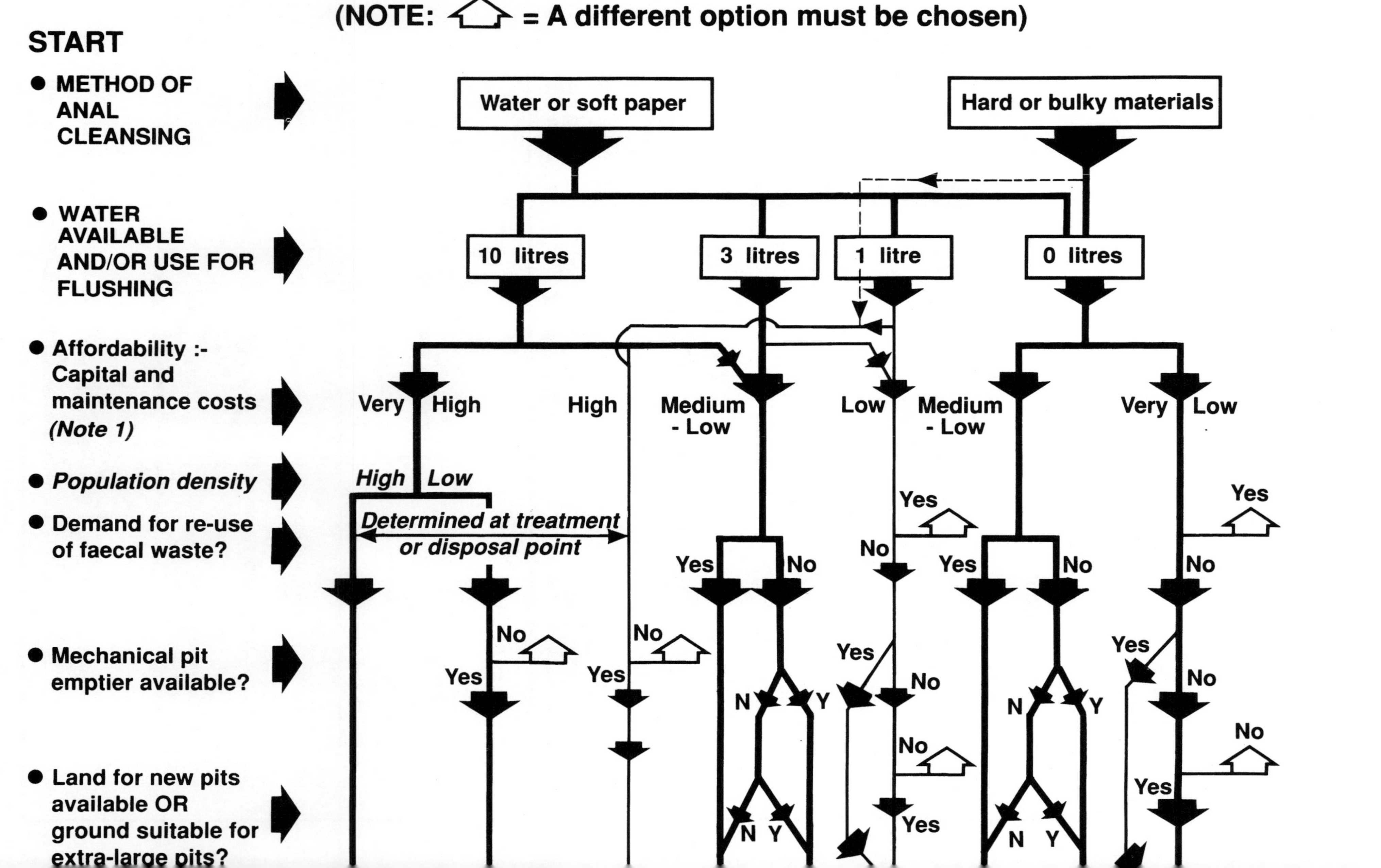

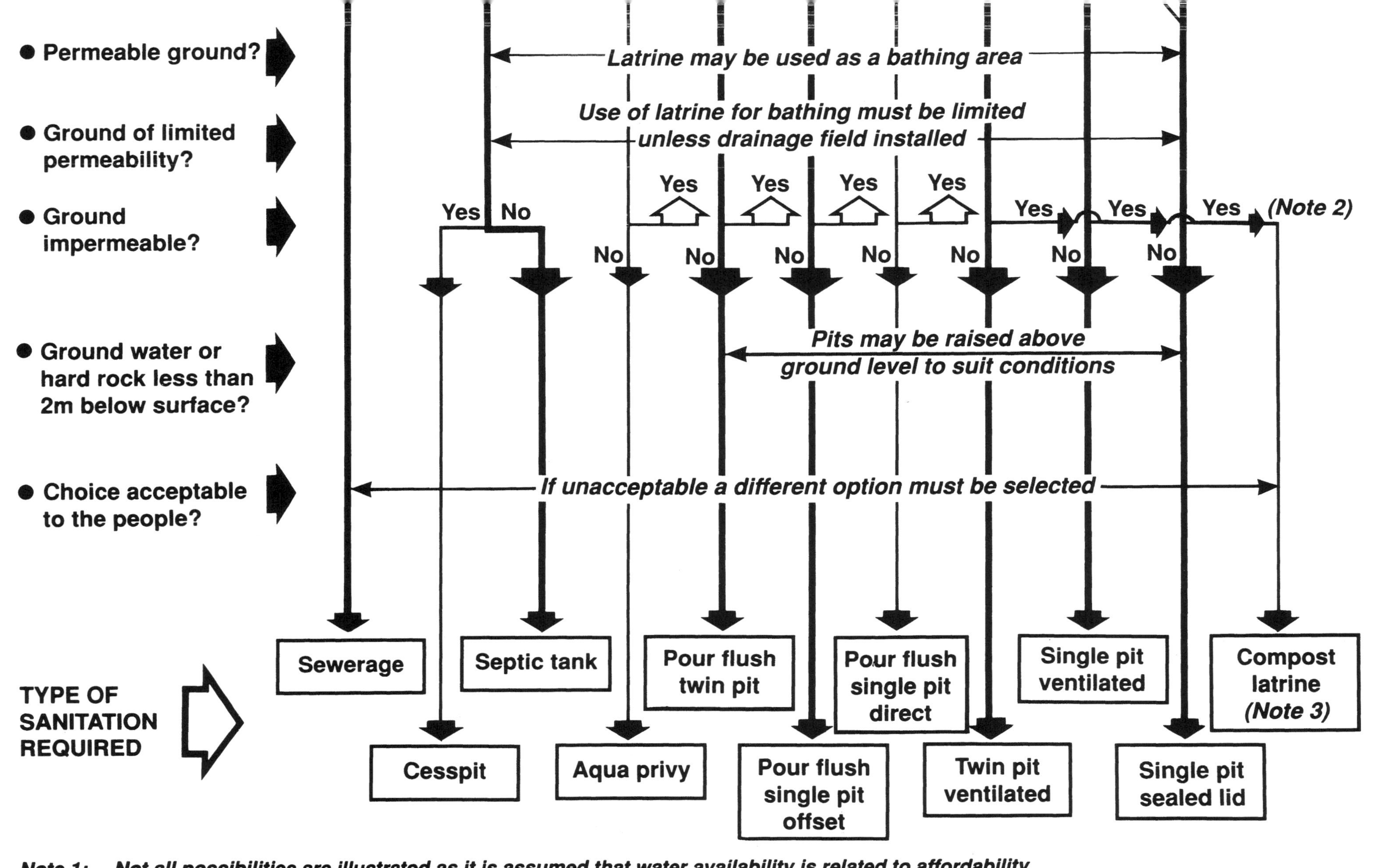

Note 1: *Not all possibilities are illustrated as it is assumed that water availability is related to affordability*

Note 2: *Use extra large pits or consider composting*

Note 3: *Also dependent on willingness to collect urine separately, demand for compost, availability of ash or vegetable matter etc.*

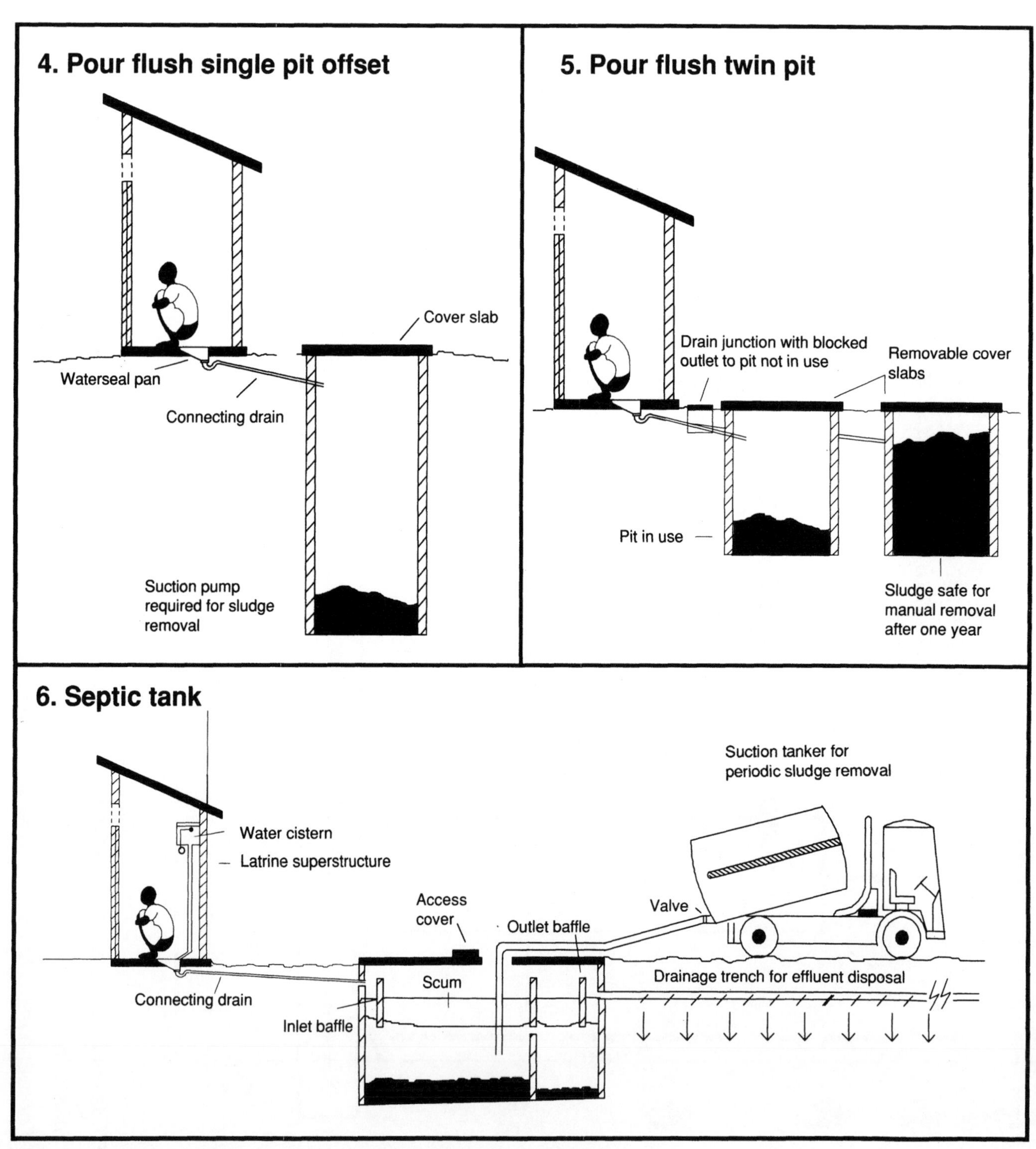

Note: In all systems, seats may be used as an alternative to squatting

For further information:

Franceys, Pickford, Reed, *On-Site Sanitation*, WHO 1991.

Text and design: Richard Franceys Illustrations: Rod Shaw
WEDC, Loughborough University of Technology, Loughborough, Leicestershire LE11 3TU, UK.

24. Groundwater dams

Groundwater dams are artificial structures that intercept or obstruct the natural flow of groundwater and provide storage for water underground.

They have been in use for many hundreds of years and are used in several parts of the world nowadays. Their use is in areas where flows of groundwater vary considerably during the course of the year, from very high flows following rain to negligible flows during the dry season.

Groundwater dams provide storage to regulate the flow of groundwater and to provide constant storage for a reliable water supply. Excess water flows over the top of the dam to replenish aquifers downstream.

There are two main types of groundwater dam:

The sub-surface dam

A sub-surface dam intercepts or obstructs the flow of an aquifer and reduces the variation of the level of the groundwater table upstream of the dam.

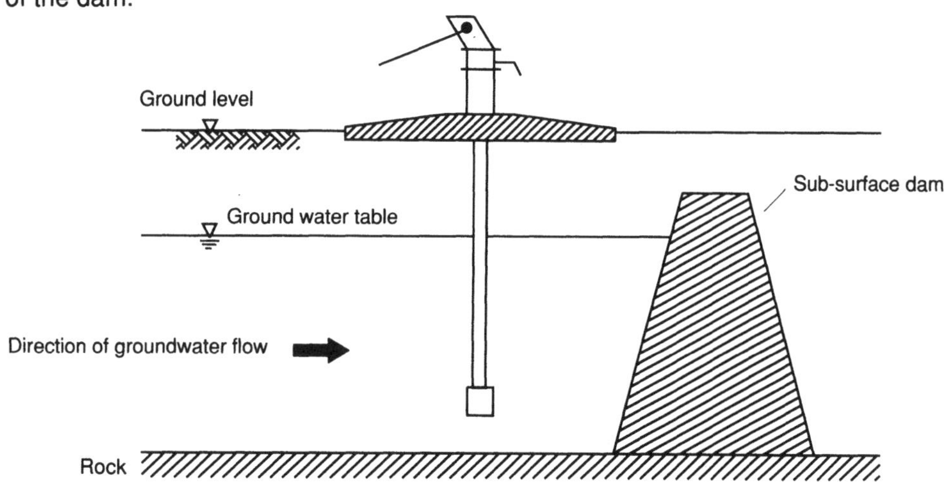

The sand storage dam

A sand storage dam obstructs the seasonal flow of water in streams. Sand and soil particles transported during periods of high flow are deposited behind the dam, and water is stored in these soil deposits.

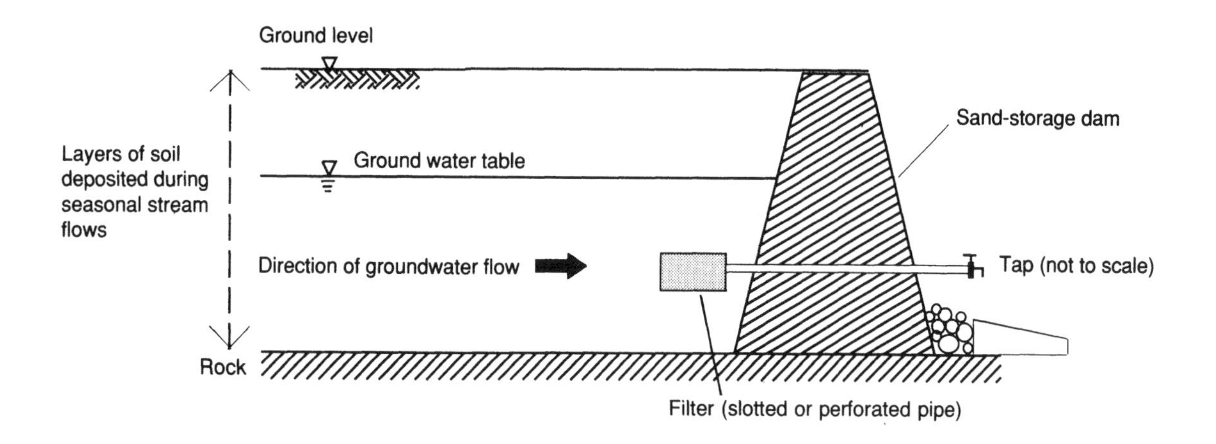

GROUNDWATER DAMS

Sub-surface dams

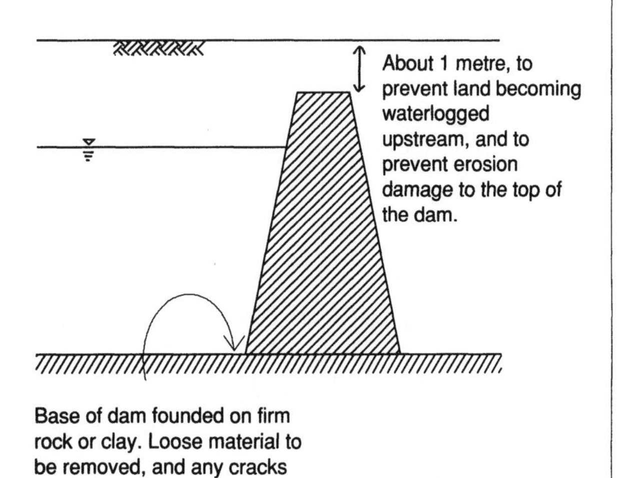

Sand storage dams

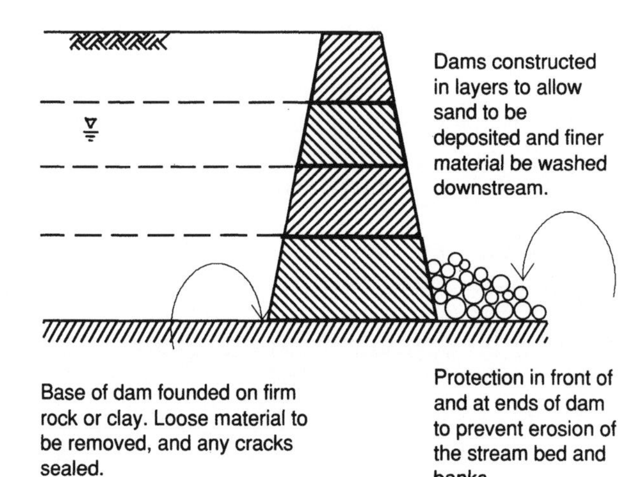

The advantages of water storage behind groundwater dams are:

- Evaporation losses are much less for water stored underground than for water stored in open reservoirs.
- The risk of contamination of the stored water by people and animals is reduced because the underground water is protected.
- A clean and reliable water source can be provided by constructing a permanent system for water collection.
- Insects and parasites (such as mosquitoes and Bilharzia parasites) cannot breed in water that is stored underground. Similarly, algae will not grow in underground reservoirs.

The full reservoir volume behind a groundwater dam is not available for storage of water, which can only be held in the spaces between soil particles. Sand is a better soil for storage than the finer silts and clays, because water flows more easily through sands, more stored water can be collected from sands, and sands reduce evaporation losses.

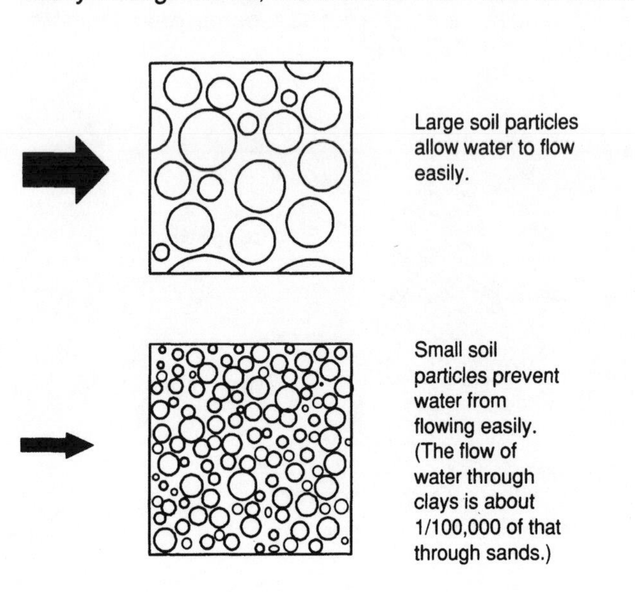

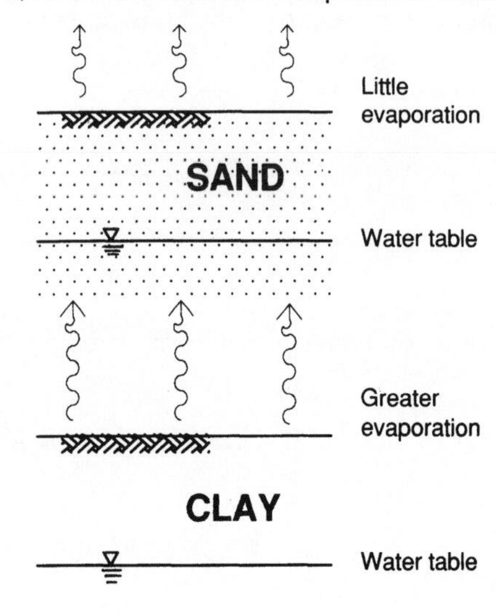

Water may be obtained from the underground reservoir either:

a) from a well upstream of the dam and preferably fitted with a handpump. (This may be used with either type of dam.)

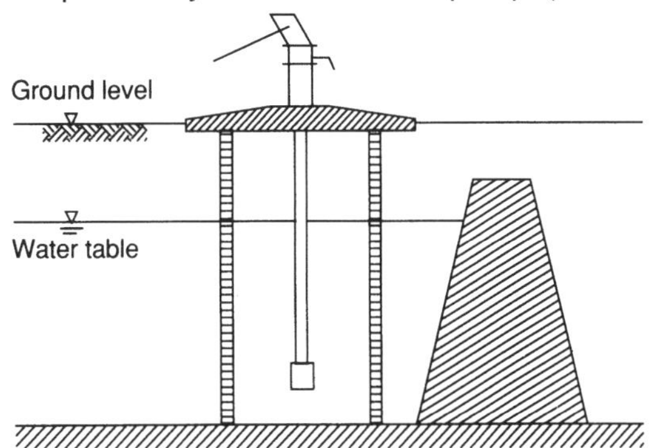

b) from a pipe, passing through the dam, and leading to a collection point downstream. (This is most suitable for sand storage dams.)

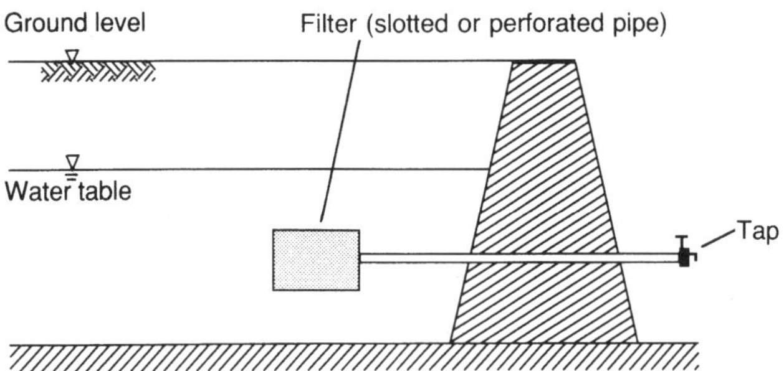

Sites for construction

The best sites for construction of groundwater dams are on gently sloping land (typically slopes of from 1:500 to 1:25) where the soil consists of sands and gravels, with rock at a depth of a few metres. Ideally the dam should be built where rainwater from a large catchment area flows through a narrow passage.

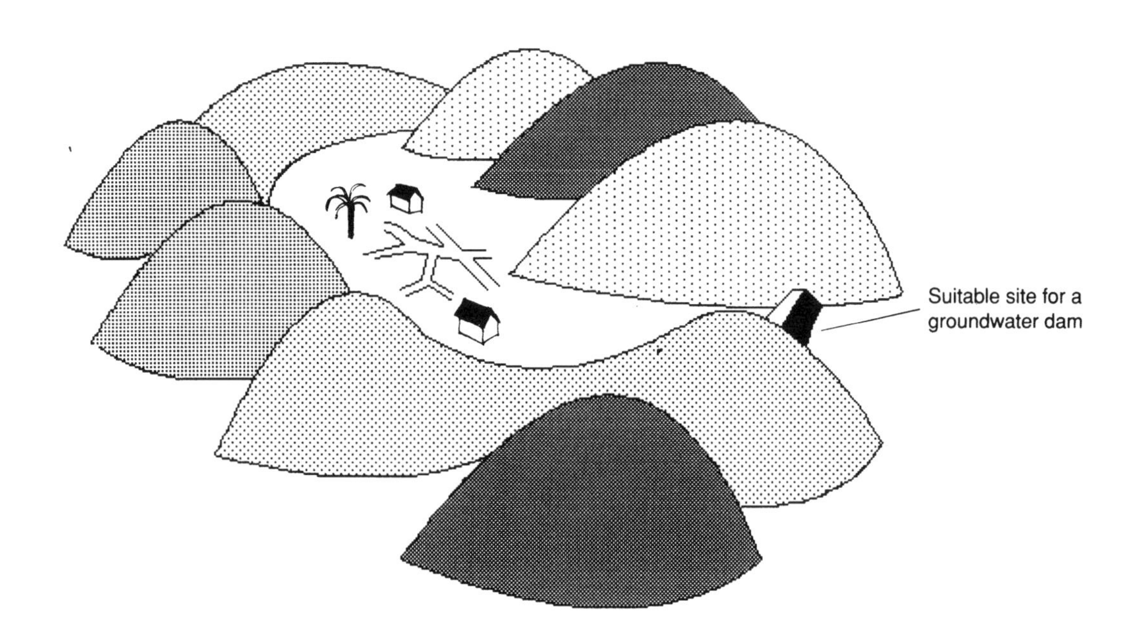

GROUNDWATER DAMS

Materials for construction

Various materials may be used for the construction of groundwater dams. Some of these are illustrated. Materials should be waterproof, and the dam must be strong enough to withstand the imposed soil and water loads. Depending upon the strength of the material used for construction, dams may be from 2 to 10 metres high. The diagrams below show materials for sub-surface dams but the first four shown could equally well be used for sand storage dams.

Warning: Dams need to be carefully designed and constructed if they are to safely carry imposed loads from soil and water without collapsing. Advice from an experienced engineer should be sought before construction starts on any dam.

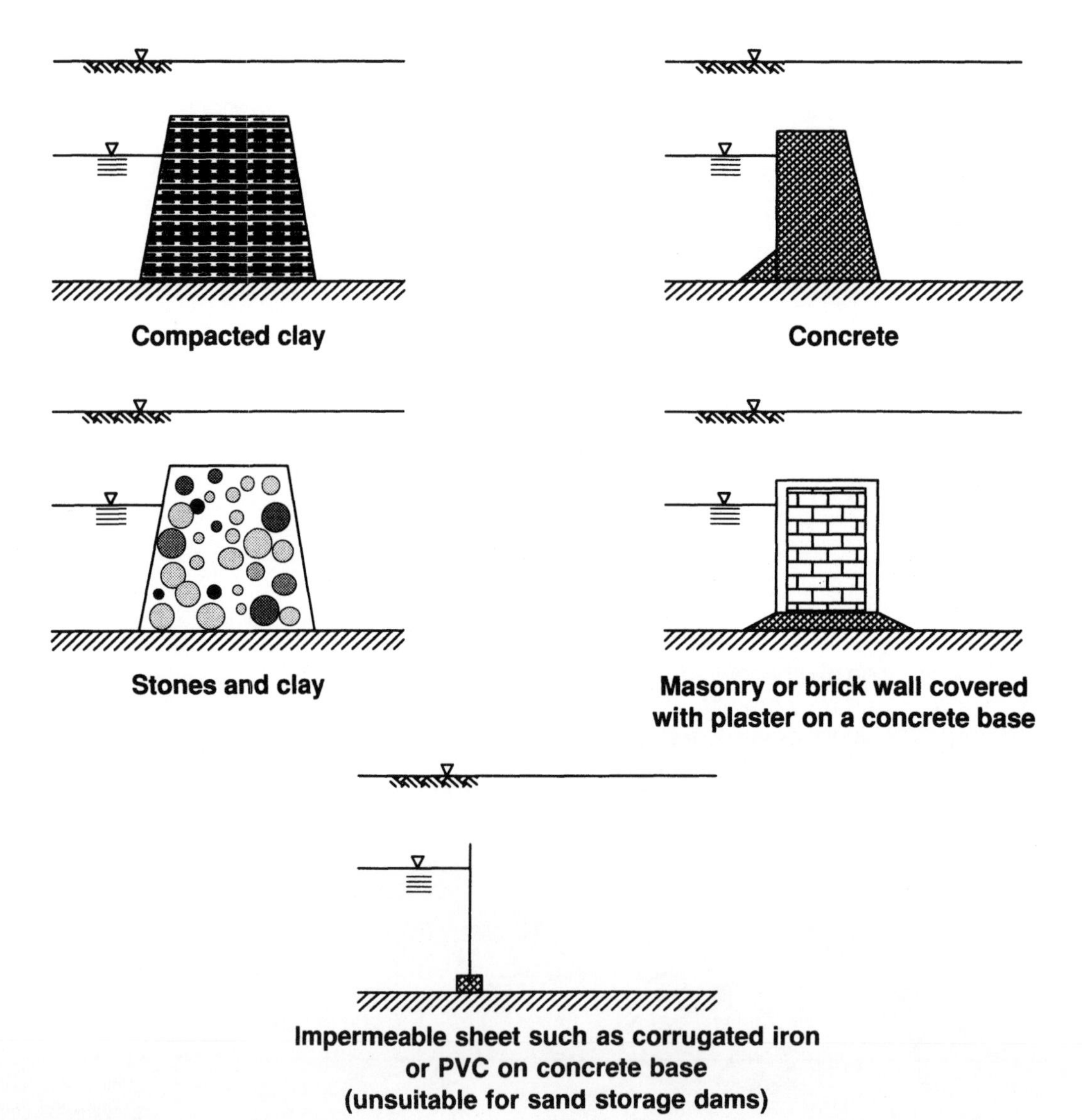

Compacted clay

Concrete

Stones and clay

Masonry or brick wall covered with plaster on a concrete base

Impermeable sheet such as corrugated iron or PVC on concrete base (unsuitable for sand storage dams)

For further information:

Nilsson, A., *Groundwater Dams for Small-scale Water Supply,* IT Publications, 1988.

Nissen-Petersen, E.,*Rain Catchment and Water Supply in Rural Africa: A Manual,* Hodder and Stoughton, 1982.

World Neighbors. *Sub-surface dams. A water catchment system built by villagers,* Filmstrip from World Neighbors, 5116 N. Portland Ave., Oklahoma City, Oklahoma 73112, USA.

Text: Michael Smith   Illustrations and design: Rod Shaw

WEDC, Loughborough University of Technology, Loughborough, Leicestershire LE11 3TU, UK.

25. Eye and skin diseases

Introduction

One of the transmission mechanisms for water-related diseases is water-washing. It is specific to those diseases dependent on water quantity but excludes those that are faecal-orally transmitted. (Technical Briefs 17 and 19, respectively). As the majority of water-washed diseases affect the skin and eyes, this Technical Brief considers the effects of both hygiene practices and the availability of water on skin and eye disease. Two notable diseases not in this group, Onchocerciasis (river blindness) and Xerophthalmia (nutritional blindness), are included because of their impact on numbers of blind people.

In tropical and subtropical developing countries skin and eye diseases are common causes for visiting a health clinic. Reduced incidence would, therefore, be beneficial to patients and staff. Some pathogenic skin and eye diseases are given in Table 1.

Table 1. Pathogenic and parasitic skin and eye diseases

Organism type	Examples of diseases/infections caused
Bacteria	Conjunctivitis *(Haemophilus aegyptius; Streptococcus pneumoniae)* trachoma *(Chlamydia trachomatis)* yaws *(Treponema pertenue)* Staphylococcal infections such as impetigo, cellulitis, boils, carbuncles etc: tropical ulcers (Vincenti's organisms)
Fungi	Ringworm (tinea or dermatophytosis) - athlete's foot (tinea pedis) - scalp ringworm (tinea capitis)
Viruses	Warts (human papilloma virus) cold sores (herpes simplex virus) conjuctivitis (picorna and adenovirus)
Parasites mites fleas worms	Allergic reaction at site of bite scabies *(Sarcoptes scabiei)* chiggers *(Tunga penetrans)* onchocerciasis *(Onchocerca volvulus)*

Eye disease

Two-thirds of the 28 million blind people in the world live in the developing countries, where blindness rates can be 10-20 times the rates in developed countries. People and particularly children under five years old living in a poor environment, with inadequate housing, sanitary facilities, food intake and health care are most at risk. By improving services and hygiene practices in these areas, up to 80% of blindness could be prevented.

The eye has its own protective mechanisms, some of which are shown in Figure 1. These are weakened by illness, poor diet, hygiene and chemical or physical damage.

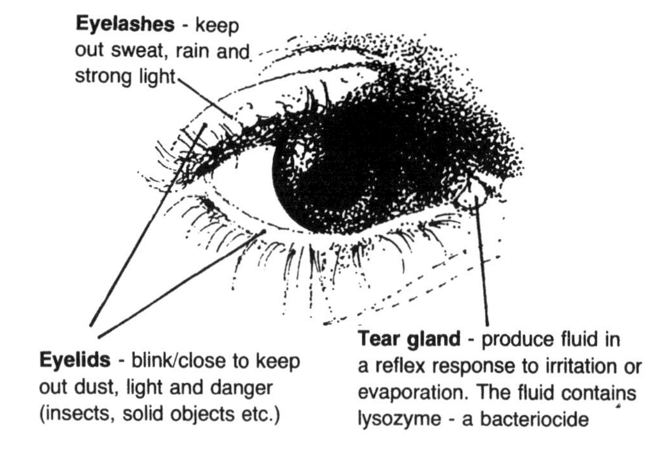

Figure 1: The eye and its protective mechanisms

EYE AND SKIN DISEASES

Conjuctivitis and trachoma, common water-washed eye diseases, are transmitted by dirty hands and towels and sometimes by flies. Trachoma affects over 500 million people, blinding seven to nine million of them through scarring of the conjunctiva, distortion of the eyelids and opacification of the cornea.

Onchocerciasis (African river blindness) results from infestation with worms *Onchocerca volvulus* which are transmitted by blackflies (*Simulium* species) when they bite. Microfilariae can cause irritation and repeated scratching damages the skin but for one million of the 20-30 million people affected microfilariae reach the eye causing permanent blindness. Because prevention through widespread use of drugs is difficult, control of the vector, by insecticide spraying is often preferred. As flies are widespread and the worm is long lived, control programmes are long-term and expensive. They are proving successful in West Africa.

Xerophthalamia (nutritional blindness), eye lesions that can result in blindness, is due to vitamin A deficiency, caused by a deficient diet or losses in repeated diarrhoeal attacks or severe illness. In Asia it affects over five million children annually, blinding 500,000; many die because of lowered resistance to other diseases. Sight is saved by early treatment with vitamin A in food (green leafy vegetables) or as supplements. Education on nutrition, therefore, is essential.

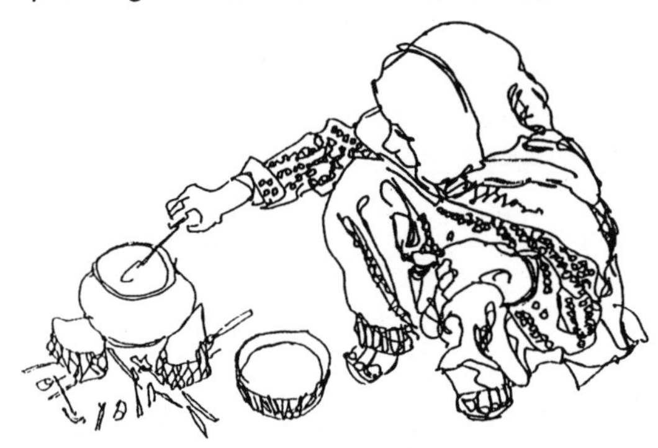

Figure 2: Education on nutrition is essential

Skin disease

Skin provides an almost continuous, waterproof and protective barrier to harmful agents, chemical and pathogenic. It assists in the control of body temperature, moisture content and waste disposal but also gives food, protection, warmth and moisture to many micro-organisms and larger parasites such as fleas and lice. Even clean, healthy skin has around five million bacteria per square centimetre, that is about 100,000,000,000 bacteria on an adult. Most of these bacteria are not harmful, in fact they may help to keep the skin clear of dead cells, sweat and, by producing acids and other chemicals, fungi. Yet, if hygiene is poor and the skin is broken or punctured (cuts, insect bites etc.), harmful agents that come into contact with it can cause disease. Resistance to disease is reduced further by illness, malnutrition or stress.

Figure 3: At least 30-40 litres of water per person per day is necessary for personal and domestic hygiene

Control of water-washed diseases

It is generally accepted that at least 30-40 litres of water per person per day is necessary for adequate personal and domestic hygiene. Even at rates greater than this, education on the benefits of hygiene practices may be needed to reduce water-washed disease. To reach this level of water availability and use, measures in the following list should be considered:

- reduce distances to water sources to less than 250 metres
- improve or increase ground and surface water sources
- supplement sources with domestic rainwater harvesting
- reduce losses by good operation and maintenance
- build washing slabs and showers
- provide soap, especially in schools
- provide hygiene education in schools, clinics, community centres
- initiate environmental improvement projects, including housing

EYE AND SKIN DISEASES

To reduce the incidence of water-washed diseases good personal hygiene practices are vital. Some of the problem areas and solutions are illustrated below.

A guide to personal hygiene

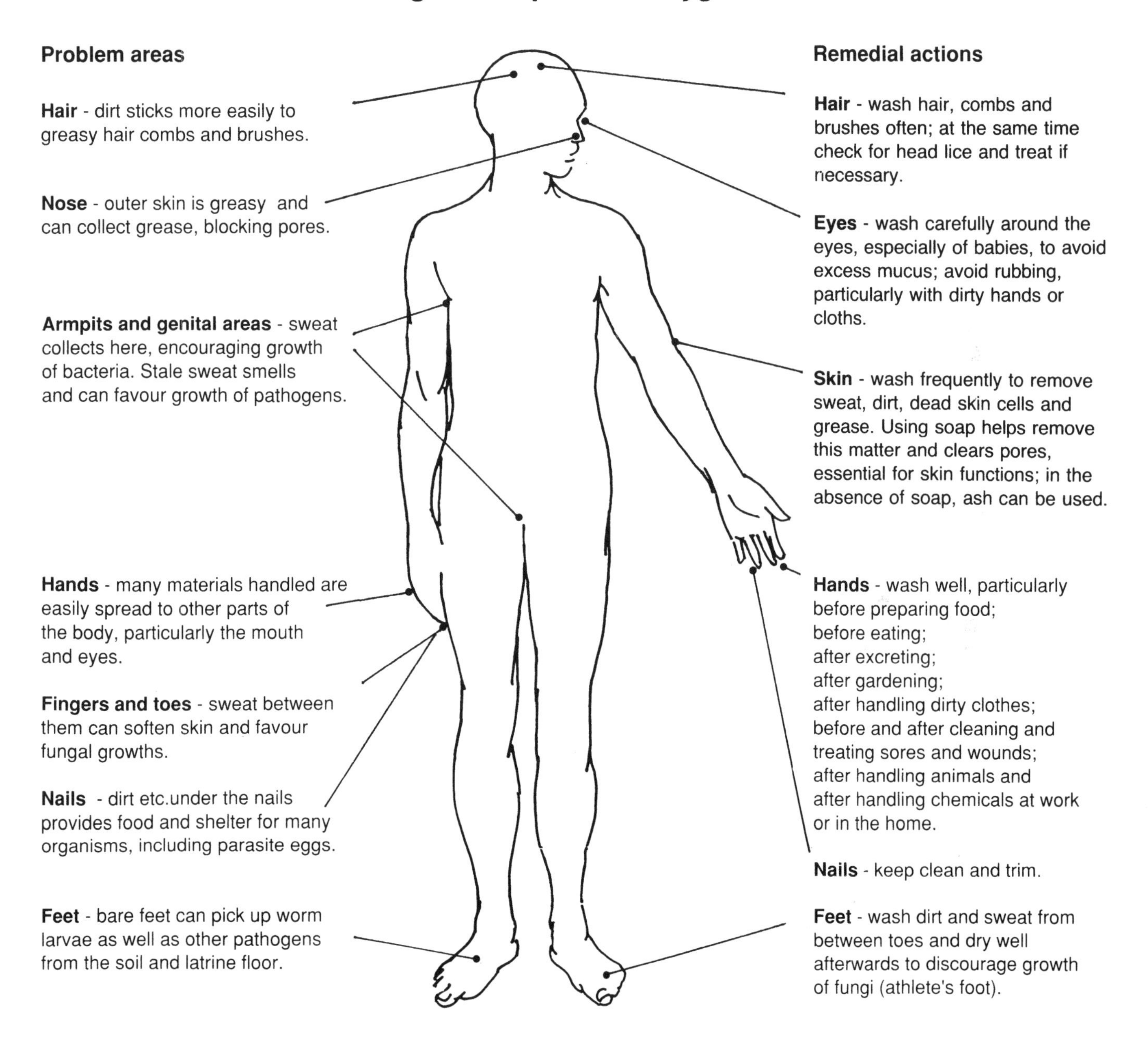

Using natural fibres, such as wool and cotton in clothes and bedding is better than using man-made fibres, such as nylon and polyesters, as they allow the skin to breathe and sweat to evaporate. Care must be taken to avoid transmission from clothes to skin of eggs laid by bot flies, such as the tumbu fly *(Cordylobia anthropophaga)*. The eggs hatch and fly larvae penetrate the skin producing large painful lesions from which the mature larvae emerge and fall to the ground. Sepsis often occurs at these exit sites. The practice of drying clothes on the ground increases transmission; ironing clothes kills the eggs.

Cleaning and washing are essential for good health, good skin and good eyes

Figure 4. Transmission of water-washed disease is decreased if houses and surrounding areas are kept clean and if bodies, hair, clothing and bedding are washed frequently.

References and further information:

Anderson, A. *World Health,* pp. 14-15, March 1986,
Oncho: a concerted effort.

Technical Briefs
- No. 8, *Making soap.*
- No. 11, *Rainwater Harvesting.*
- No. 14, *Above-ground rainwater storage.*
- No. 17, *Health, water and sanitation I.*
- No. 19, *Health, water and sanitation II.*

Thylefors, B. *WHO Chronicle,* Vol.39, No. 4, pp149-54, 1985,
Prevention of blindness: the current focus.

Truswell, A. S. *British Medical Journal* Vol.291, pp587-89, 1985,
Malnutrition in the Third World II.

World Health,
A decade of oncho control, October 1985.
Health for all - all for health, January-February 1988.

Text: Dr Margaret Ince   Illustrations and design: Rod Shaw
WEDC, Loughborough University of Technology, Loughborough, Leicestershire LE11 3TU, UK.

26. Public standposts

Public standposts provide points where a local community may draw water from a piped water distribution system. They usually comprise a connection to the water main, a suitably supported riser pipe and a tap. Their design and construction has a major influence on their durability, effectiveness and hygiene. However, standposts often receive inadequate attention and failures are frequent. This affects many people, both in rural and urban areas, as standposts often represent the only feasible and affordable means of access to water.

A well-designed standpost must:

- provide sufficient quantities of water to all users when it is needed;
- be durable and reflect local customs;
- contribute towards improvements in public health.

Water use and location

The amount of water used per person depends upon walking distance to the standpost. The standpost must also be able to meet the water requirements of the community served during times of high demand (usually in the early morning and early evening, covering about six hours).

Distance to source:	Not exceeding 250m (<200m where possible)
Water usage:	20-60 litres per person per day depending upon distance
Population served:	150-250 persons per standpost and up to 125 persons per tap

A minimum of 20 litres per person per day (but ideally 35 litres per person per day) should be assured to achieve the benefits of an improved water supply.

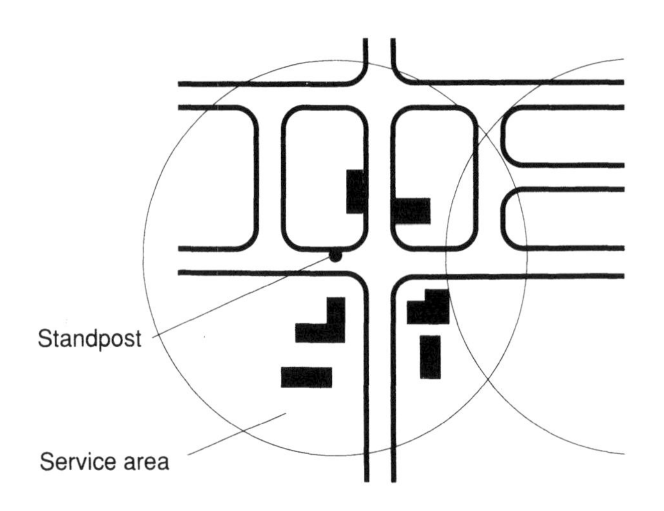

Good access is important both for users and maintenance. The standpost should be located centrally in the area it serves, ideally at a road junction, and must be on public land to avoid access problems. In hilly areas, remember it is easier to carry full containers downhill rather than uphill. Easy removal of waste water must also be considered.

Coping with peak demands

An adequate flow of water must be available during the hours of peak demand. The required discharge (in litres per hour) is given by:

$$Q = \frac{P\,C\,(1+w)}{H}$$

where: Q is the flow of water required in litres per hour;
P is the population served (making allowance for any growth during the life of the standpost);
C is average water usage in litres per person per day, including any use of water at the tap (bathing or laundry etc.). Account must also be made of any water used for other purposes (irrigation or animals etc.);
w is an allowance for wasted water (typically 10-40% of water usage);
H is the number of hours in the day that the standpost is in constant use (6-12 hours typically).

Typical figures for this calculation might be:

$$Q = \frac{P\,C\,(1+w)}{H} = \frac{200 \times 35\,(1+0.30)}{6} =$$

1 517 litres per hour
0.42 litres per second

A 12mm standard tap will discharge about 0.22 litres per second under normal pressure, and in the example above, a standpost provided with two 12mm tap outlets would suffice. Alternatively, a 19mm tap will discharge 0.42 litres per second under similar conditions. These are the minimum requirements. Additional taps could be considered to reduce queuing during periods of peak demand.

Hydraulic design

As water flows from the main and through the standpost, pressure is lost for three reasons:

- **Change in elevation:** invariably, the taps will be higher than the water main and available pressure is reduced by their difference in elevation.
- **Pipe friction:** energy losses depend upon the length of pipe, its diameter, and rate of flow. The table below allows approximate values for losses to be calculated. Fittings, such as bends, tees etc. cause additional losses and a rough allowance for these losses through a typical standpost layout may be obtained by increasing the pipe length by 10%.
- **Taps:** major losses occur as water passes through a tap depending upon rate of flow and the tap design. Head losses of around 2.0m are typical.

Flow	Size of GI pipe (mm)		
(l/hr)	12	19	25
500	0.30	0.03	-
750	0.65	0.06	0.01
1 000	1.10	0.10	0.03
1 500	-	0.23	0.06
2 500	-	0.65	0.15

Approximate head losses due to pipe friction, expressed in metres per metre length for galvanized pipe.

To ensure that an adequate flow of water is available at the tap, the total head lost must equal or be less than the head available in the main at the point of connection.

Total head loss	=	diff. in elev. of main and tap	+	loss in pipes and fittings	+	loss through the tap	<	head in main

In critical situations where the energy losses will exceed the head available in the main, they may be reduced by shortening the connecting pipe; increasing its diameter; relocating the standpost to reduce the difference in elevation between the tap and the main; or increasing the size and number of taps provided. Where considerable surplus energy is available, a valve may be added to restrict the flow and dissipate energy.

Layout and details

Public standposts are subject to heavy use and abuse. They are also the contact point between the public and the water supply. They must therefore be simple to construct, durable and easy to maintain, and reflect local social and cultural needs.

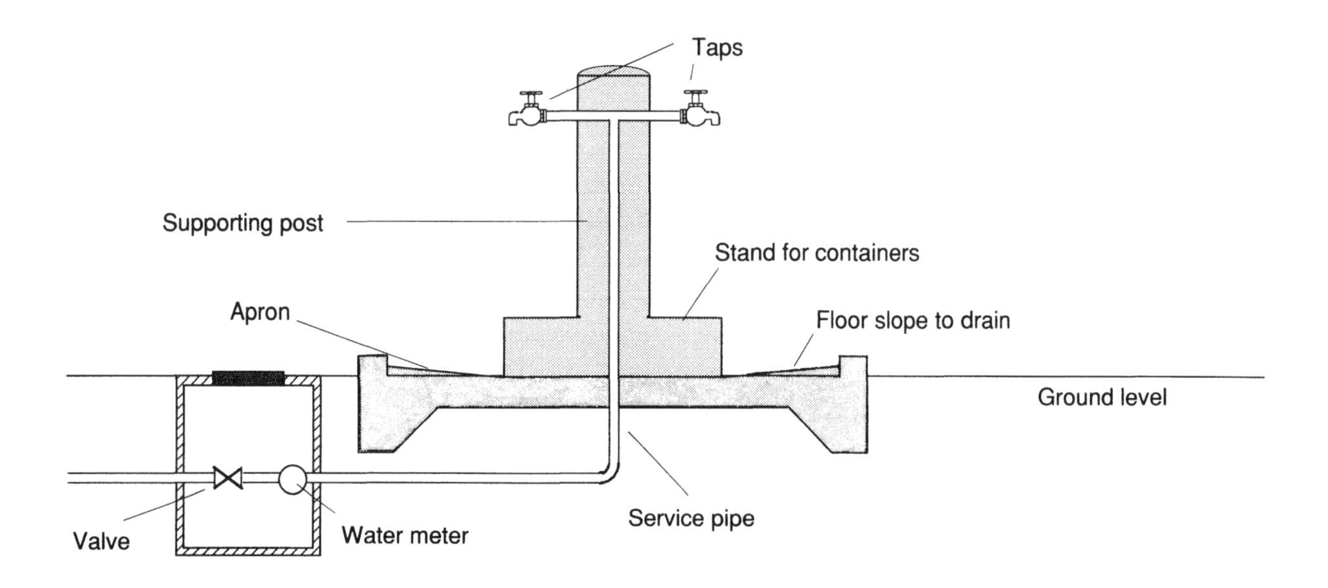

The main components of the standpost are:

The supporting post
The supporting post encases the riser pipe and is made of durable materials such as concrete or masonry. Ideally it is about 300mm square and extends 100mm above the taps to protect them. A raised stand under the taps may be added to support containers while being filled, depending upon local customs.

The platform or apron
The platform or apron extends at least one metre all around the taps. Where bathing or laundry is carried out at the tap, the apron should be extended to collect all waste water. It may be made of concrete at least 150mm thick and lightly reinforced to prevent cracking. An upstand around the perimeter will contain the waste water and a floor slope of between one in 50 and one in 100 will direct the waste water to the drain outlet.

The service pipe
The service pipe may be galvanized iron (or PVC if it is well protected), of diameter 12mm to 36mm depending upon the number of taps served. A main valve is required to isolate the unit, and a further control valve can be incorporated if flows need to be restricted. A meter may be used for charging for consumption and monitoring, but it is often prone to damage and must be well secured in a lockable box together with any control devices.

The taps
The taps should be robust and easy to maintain. They constitute a small proportion of the total standpost cost but are a major source of problems. Hence, careful selection is necessary and the best quality that can be afforded should be fitted. Spring loaded taps will reduce wastage but are frequently broken by users. The height of the taps should be convenient for the users (often women and children) and is typically 0.7m to 1.0m.

Drainage

The disposal system for waste water from the standpost area is an integral part of standpost design. Waste water should be prevented from running on to surrounding ground and should be directed to gutters, lined drains or natural drainage channels where possible. An alternative is the construction of a soakaway pit similar to that used for a septic tank. Its size will depend upon the soil permeability but is typically 0.5m^2 in area and 0.8m deep, filled with rubble or gravel through which the waste water can percolate into the ground. Crop irrigation may also be considered, provided the crops need water all the year round. In difficult situations a larger number of single tap standposts may be considered to spread the waste water disposal over a wider area.

All these factors must be considered when designing a public standpost in order to provide a water source that can satisfactorily cope with the heavy demands likely to be placed upon it.

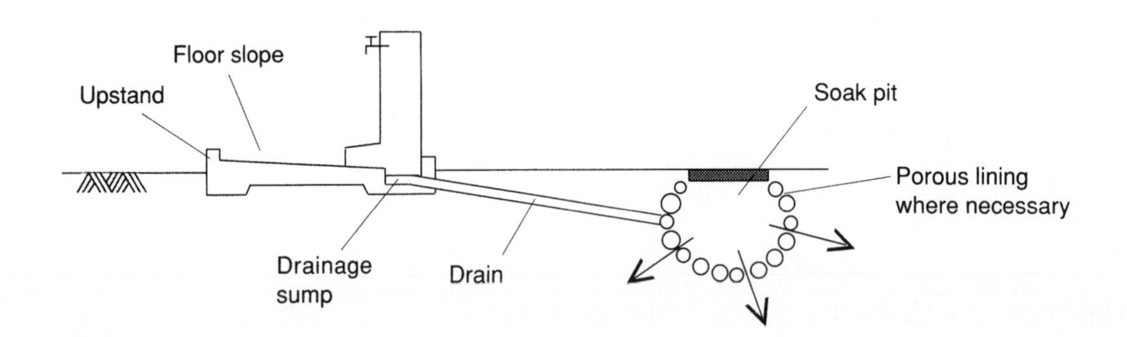

For further information:
IRC, *Public Standpost Water Supplies, A Design Manual,* Technical Paper No. 14, WHO International Reference Centre for Community Water Supply, The Hague, Netherlands, 1979.
Technical Brief No. 7: *The Water Cycle*, and Technical Brief No. 12: *Septic Tank and Aquaprivies*.

Text: Alistair Wray Illustrations and design: Rod Shaw
WEDC, Loughborough University of Technology, Loughborough, Leicestershire LE11 3TU, UK.

27. Discharge measurements and estimates

Introduction

Community water-supply schemes are generally designed for water consumption in the approximate range of 15 to 60 litres per person per day. Allowance has also to be made for water use by livestock, for future growth in population and demand, and for losses of water by leakage etc. from the system.

It is important to check that the discharge of the source is enough, throughout the year, to meet the community's requirements. This requires measuring or estimating the dry season flow, when least water will be available.

If a small reservoir is to be built, more detailed flow data are needed:

- Year round flows, to size the storage
- Flood flows, to design the spillway

Useful conversion factors

One litre per second	= 1 l/s	One cubic metre	= 1000 litres
	= 86 400 litres per day	One Imp gallon	= 4.546 litres
		One US gallon	= 3.785 litres

Ways to measure the flows of springs and small streams

Bucket and stopwatch

In this method, all the flow from a spring or small stream is collected in a container whose volume is known (e.g. a bucket, jerrycan, 200-litre drum) and the time to fill the container is measured. The filling time should be more than five seconds, to give reasonable accuracy.

Discharge or yield (l/s) = volume (litres) / time (seconds)

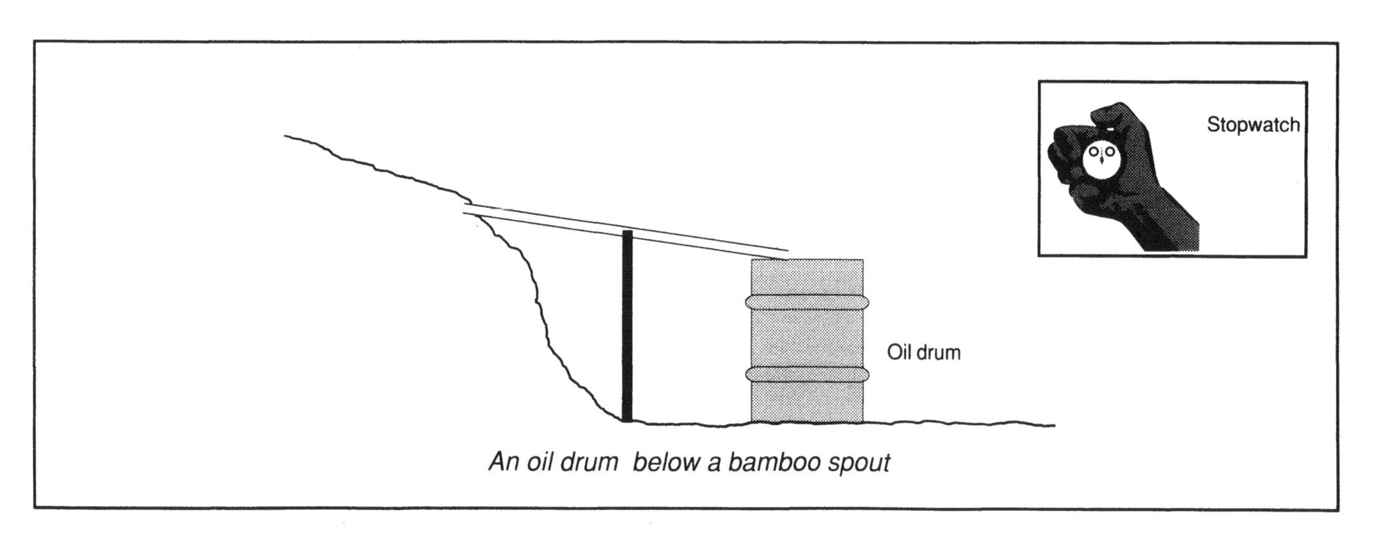

An oil drum below a bamboo spout

DISCHARGE MEASUREMENTS AND ESTIMATES

Float, stopwatch and tape

Floats are a simple way of measuring the velocity of a stream, but they are not very accurate. The surface velocity is obtained by measuring the time (t secs) for a float to travel a measured distance (L metres). It is best to choose a straight, uniform river section about 30m long, and to time the float over a number of repeated runs. A piece of fruit makes a good float, as it is less affected by wind than a wooden stick. A factor of about 0.85 should be used to convert surface velocity to average velocity.

Surface velocity (m/s) $= L / t$

Average velocity (m/s) $= 0.85 \times L / t$

The cross section of the stream should be measured up carefully in a number of places along the test distance, and the average cross-sectional area calculated (A sq m).

Discharge (cubic metres per second)

= average velocity x cross sectional area of stream

$= 0.85 \times (L / t) \times A$

Discharge (l /s) $= 1000 \times 0.85 \times (L / t) \times A$

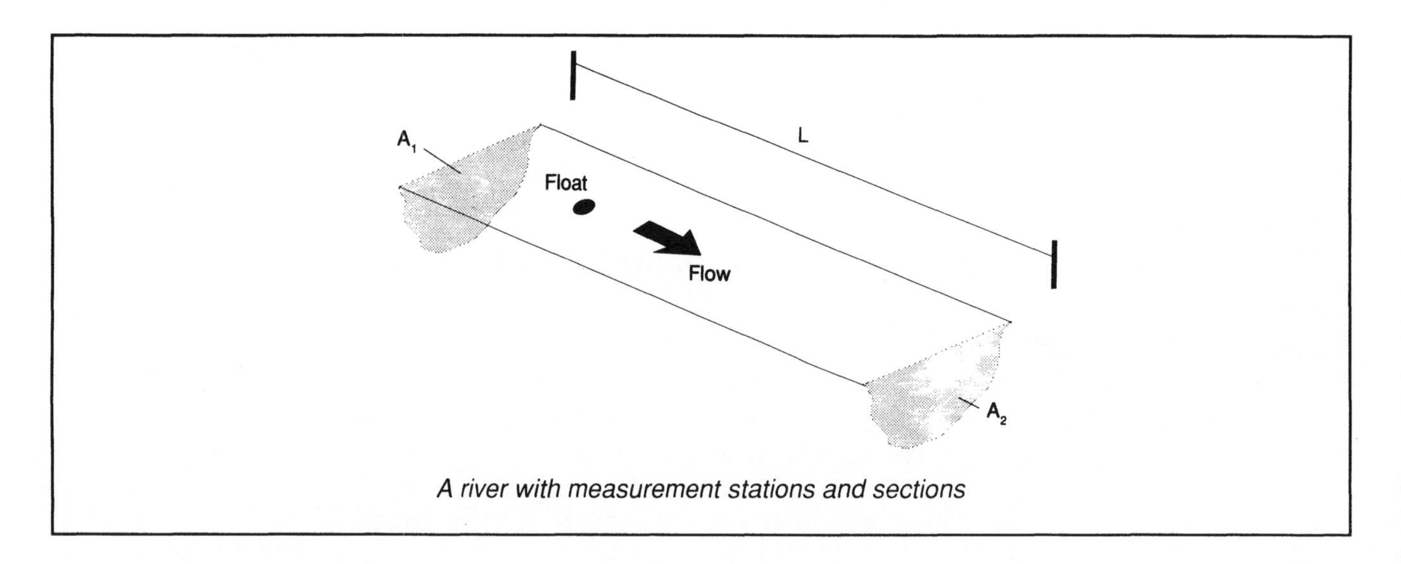

A river with measurement stations and sections

Weir with stick gauge

Portable weirs made of steel plate can be used for measuring the flow of small streams or springs. They can give accurate measurements, if they are installed carefully. The weir should be set vertically, perpendicular to the stream, and with the crest horizontal. A free fall is required over the weir crest. Leaking must be prevented around the sides of the weir, possibly by using a polythene sheet.

A stick gauge, marked in centimetres, is set vertically at the side of the stream upstream of the weir. It is used to measure the head, h (m), which is the difference between the upstream water level and the crest of the weir. It is important to check that the gauge zero is truly at weir-crest height. This may be done using a spirit level and string line, or water-filled flexible plastic tubing.

There are different shapes of weir, each of which has a standard formula for calculation of discharge. Details for two common weir shapes are given opposite.

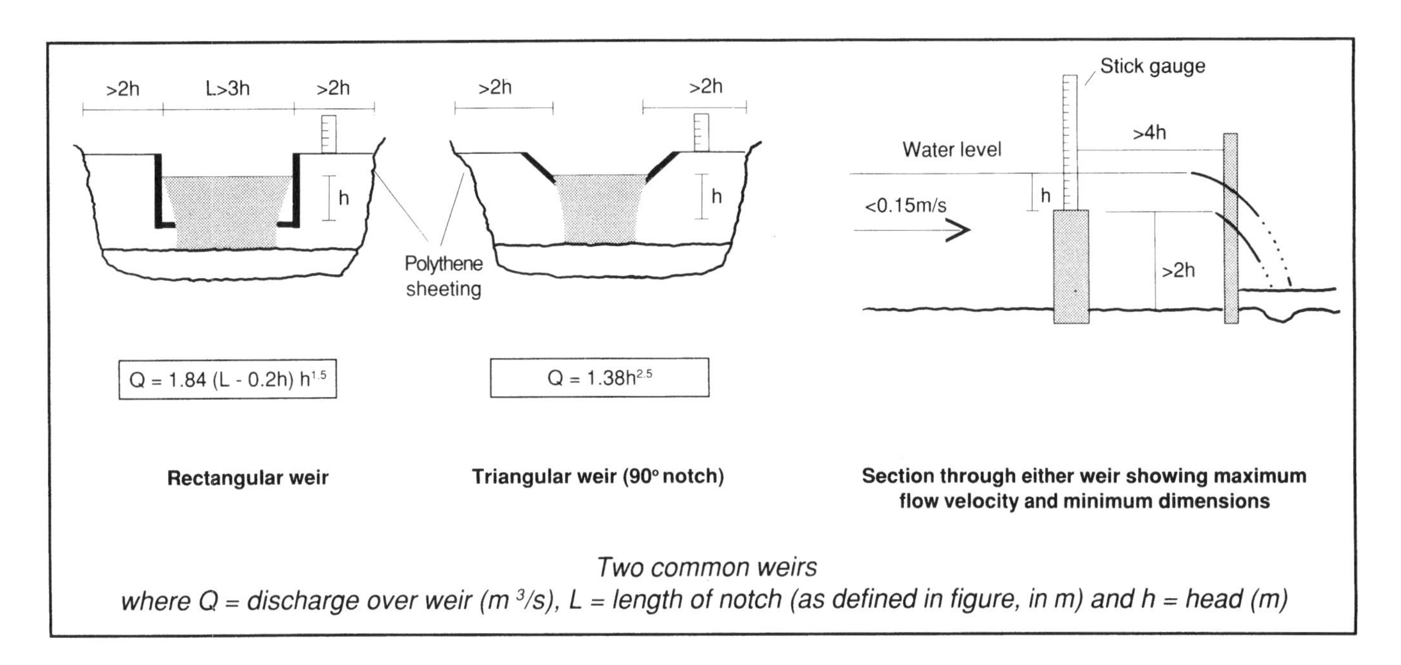

Two common weirs
where Q = discharge over weir (m^3/s), L = length of notch (as defined in figure, in m) and h = head (m)

Estimating flows of small streams

Possible sources of information for estimating the range of flows are:

a) local knowledge
b) observation and survey
c) rainfall records
d) size of catchment area
e) spot measurement of discharge

It is important to take account of other users, both upstream and downstream.

Possible ways of estimating flows in the dry season are:

a) From spot measurements during the dry season. This is the best way.
b) From spot measurements at other times of year,
either, reduced by suitable factors derived from experience on similar river basins
or, plotted against time on a graph, and extrapolated to the dry season. Records from measurement on similar river basins can be used to derive a typical shape for the graph.
c) From any records of flows in a similar river basin nearby, with similar rainfall, by using the following method:

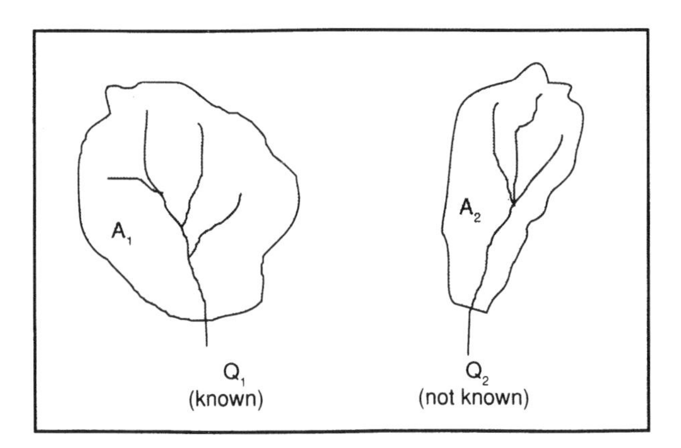

- for the recorded basin, measure the catchment area A1 which drains to the measurement station;
- for the unrecorded basin measure the catchment area A2 which drains to the point where the flow is to be estimated;
- obtain the dry season flow Q1 for the recorded basin;
- **the dry season flow Q2 for the unrecorded basin can then be estimated as: Q2 = Q1 × A2/A1.**

d) From local reports (or records) of dry season depth, coupled with survey of the cross section and slope of the stream, and using the Manning equation:

$$Q = \frac{A R^{2/3} s^{1/2}}{n}$$

Q = discharge (m^3/s)
A = area (m^2)
R = A/P where P = wetted perimeter (m)
s = slope (m/m)
n = Mannings roughness coefficient

suitable values of n are:	
straight river with earth bed:	0.02 - 0.025
straight river with stony bed:	0.03 - 0.04
winding river with earth bed:	0.03 - 0.05
winding river with stony bed:	0.04 - 0.08

Constant discharges from boreholes or pipes

These may be measured by one of the following methods:

1. Using a V-notch weir set in a steel tank or earth channel
2. Using a calibrated orifice plate and pressure gauge on an upstream tapping (the pressure difference across an orifice plate will depend on the flow rate)
3. Measuring the co-ordinates of the water trajectory from a horizontal pipe, as shown below.

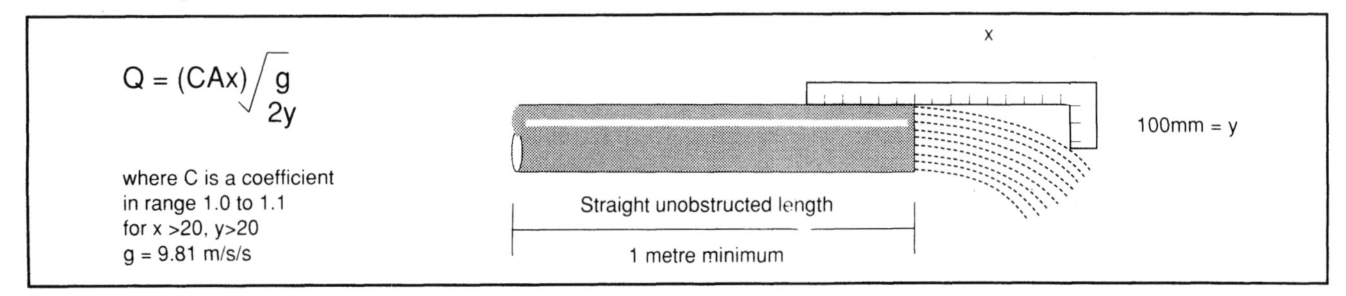

Measurement of flow from a horizontal pipe

Horizontal distance, x (mm) for fall in water trajectory y= 100mm. - See above figure	Nominal pipe diameter D (mm)				
	25	32	37	50	62
100	0.36	0.6	0.8	1.4	2.0
125	0.45	0.8	1.0	1.7	2.5
150	0.54	0.9	1.3	2.1	3.0
175	0.6	1.1	1.5	2.4	3.5
200	0.7	1.2	1.7	2.8	3.9
225	0.8	1.4	1.9	3.1	4.4
250	0.9	1.5	2.1	3.5	4.9
275	1.0	1.7	2.3	3.8	5.4
300	1.1	1.8	2.5	4.2	5.9
325	1.2	2.0	2.7	4.5	6.4
350	1.3	2.1	2.9	4.9	6.9
375	1.3	2.3	3.1	5.2	7.4
400	1.4	2.5	3.3	5.5	7.9

Water flow from horizontal pipes (litres/second) flowing completely full at the end

Further information:

Cairncross, S. and Feachem, R., *Small water supplies,* Ross Institute Bulletin No. 10, 1978.
Herschy, R. W., *Streamflow measurement,* Elsevier, 1985.

Text: Ian Smout   Illustrations and design: Rod Shaw
WEDC, Loughborough University of Technology, Loughborough, Leicestershire LE11 3TU, UK.

28. Public and communal latrines

Public and communal latrines

Public latrines are located at markets, lorry parks, bus stations and similar places where they are used by people who are away from their homes.

Communal latrines are located in or near housing areas, and are used by the community - people living in nearby houses who have no household latrines.

- **Design and construction of public and communal latrines is straightforward**
- **Operation and maintenance are usually difficult**

Therefore . . .

before starting to build a public or communal latrine there must be clear plans . . .

- for cleaning and operation
- for the wages of any paid staff
- for paying any water and electricity charges
- for maintenance and repairs
- for proper management and control

Design and construction

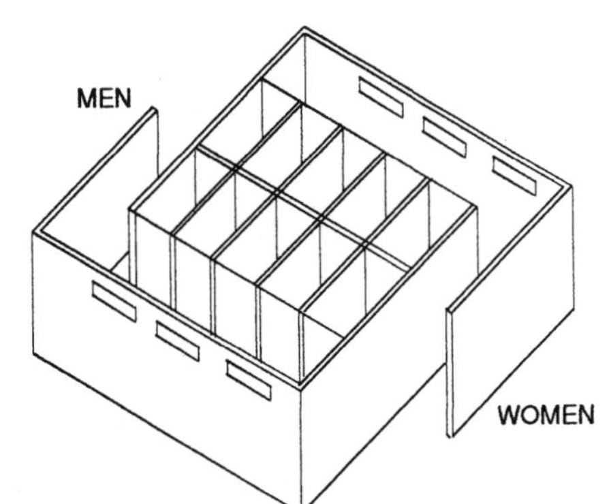

Separate facilities for men and women should always be provided.

Exceptionally, a single special toilet for disabled people (those using wheel-chairs) may be used by men and women.

Public or communal latrines are usually of the same type as the household toilets nearby.

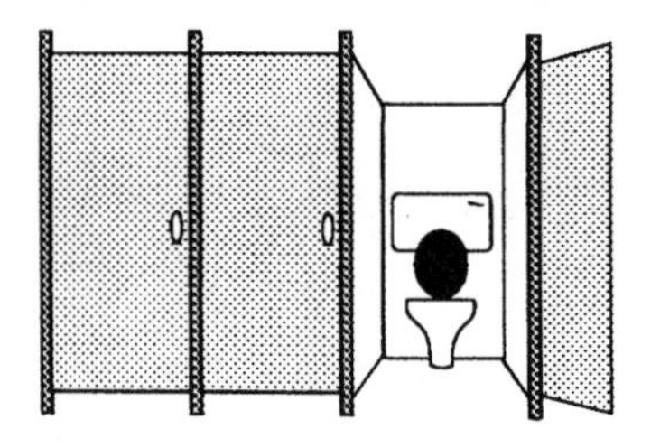

Where an ample and reliable piped water supply is available, WATER CLOSETS are appropriate.

Water closets may be connected to sewers, or to septic tanks where there is no sewerage.

The capacity for a septic tank should be sufficient for at least one day's retention of liquid, plus space for accumulation of sludge and scum.

For example, assuming: half the volume of the tank is allowed for sludge and scum; the latrine is expected to be used 1000 times each day; each WC flush uses 10 litres of water.

The tank volume required is 2 x 1000 x 10 litres = 20 cubic metres.
For a depth of two metres, the area required is 10 square metres

For a two-compartment tank, the dimensions could be:

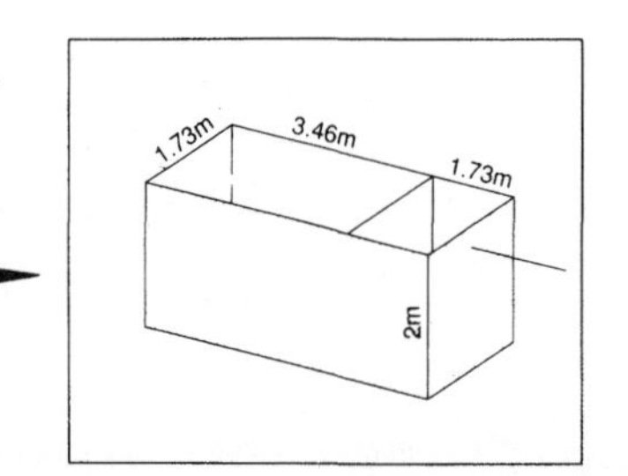

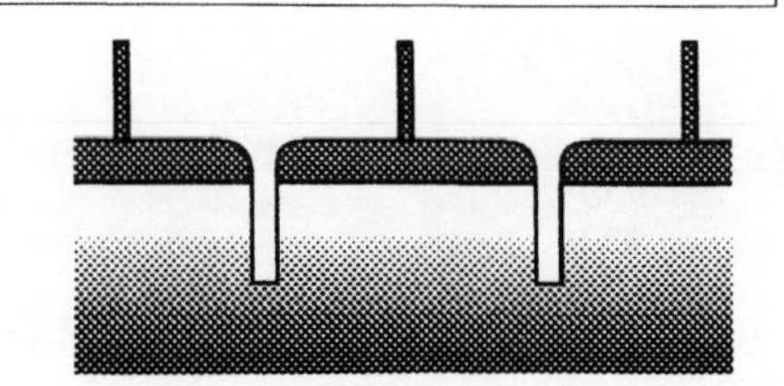

An **aqua privy latrine** (sometimes called a 'septic tank latrine') has the latrine above the tank. A vertical pipe is set below each toilet and extends below the water surface.

This type of latrine is suitable where water supply is limited or where solid material is used for anal cleansing.

Where water has to be carried to the latrine (for example, from a public standpost) facilities for bathing and/or washing clothes may be provided to keep a flow of water passing through the tank.

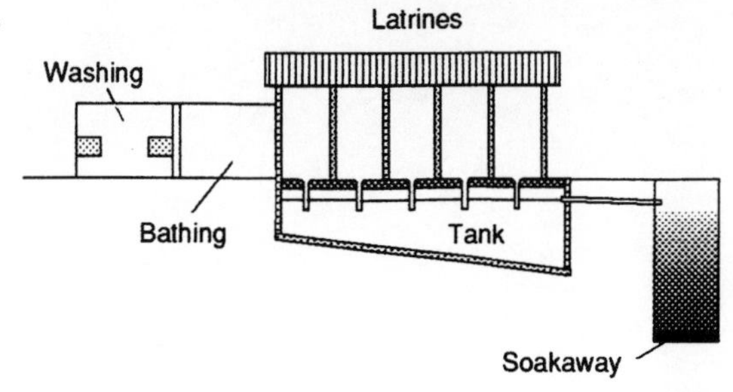

Pit latrines

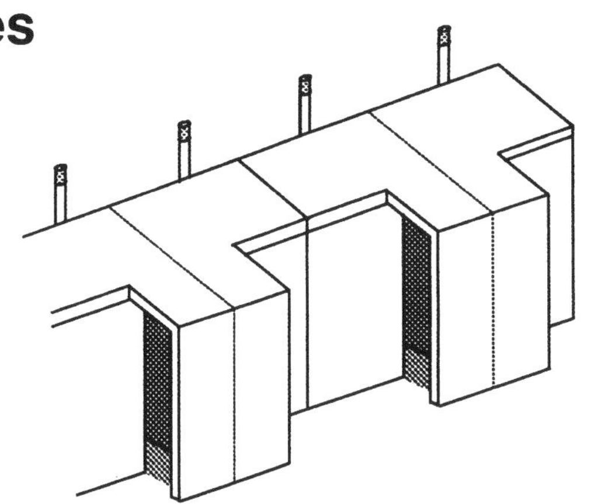

Either:

A VENTILATED IMPROVED PIT LATRINE (VIP) with darkened interior for the toilet cubicles, and adequate vent pipe (see note below) with fly proof netting.

Or:

A POUR- FLUSH LATRINE with a trap below each toilet bowl providing a 20mm deep water seal.

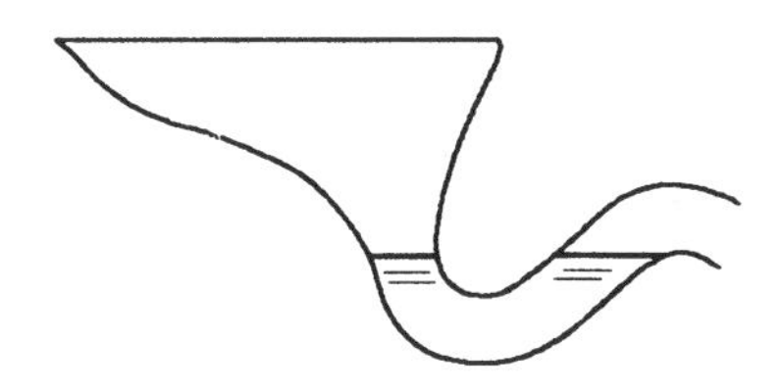

Or:

ALTERNATING TWIN-PIT LATRINES for example, the K-VIP shown here.

The number of pits or pit compartments should be one more than the number of toilet cubicles.

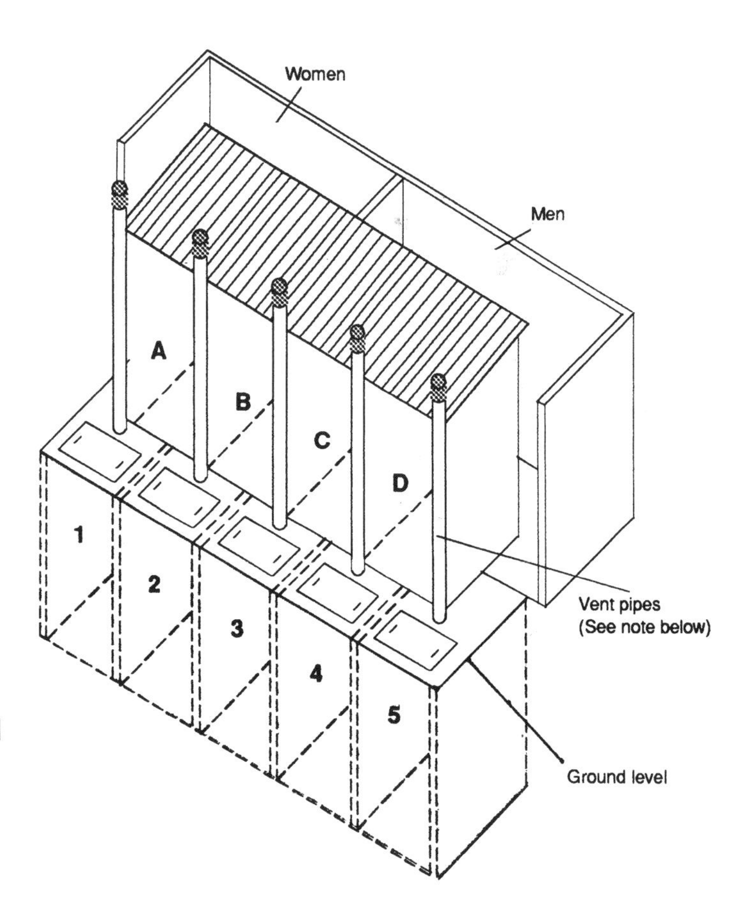

For the first two years or so:
- Toilet A goes to compartment 1
- Toilets B and C go to compartment 3
- Toilet D goes to compartment 5.

For the next two years or so:
- Toilets A and B go to compartment 2
- Toilets C and D go to compartment 4.

Then compartments 1,3 and 5 are emptied and used again.

Note:

Vent pipes should be at least: *150mm diameter if a smooth pipe (e.g. PVC)*
225mm diameter if a rough pipe (e.g. mud)
225mm square if made of blocks or bricks

Cleaning

The most important operation for a public or communal latrine is keeping it clean

Cleaning is sometimes shared by all the people who use the latrine, working to some kind of rota. **Such arrangements are rarely satisfactory.**

Cleaning is usually undertaken by people paid to do it.
Payment of wages (e.g. by a local council) must be agreed BEFORE the latrine is built.

A system which works well in some places is for users to make a small payment to use the latrine. Sometimes no charge is made for women and/or children, and/or unemployed people.

In some places, cleaning is 'privatized'. The cleaners keep the fee, pay for the water and electricity, and maintain a high standard of cleanliness because the cleaner the latrine, the more people will pay to use it.

Payment for water and electricity charges and for maintenance and repairs must be agreed BEFORE the latrine is built.

Maintenance and repairs

- **Emptying twin-pit latrines**
- **Regular inspection, and repair or replacement if necessary, of fly proof netting of VIP latrine vent pipes; and floors, traps, walls, doors and roofs; and any water and electricity equipment.**

Text: John Pickford Graphics: Rod Shaw
WEDC, Loughborough University of Technology, Loughborough, Leicestershire LE11 3TU, UK.

29. Designing simple pipelines

Introduction

This paper tries to introduce the reader to the principles of pipeline design. It is a very difficult process for people who have not previously been taught the principles of hydraulics to understand and so it has been necessary to simplify the calculations by excluding factors that usually have only a small effect on the design.

Terminology

Water flowing along a pipe is a bit like a car travelling along a road. For the car to move it must use energy. That energy can be supplied by the fuel in the fuel tank being burned and by the height of the car above the point it is trying to reach. As the car travels along some of the energy will be used up overcoming the friction in the car's moving parts. On a flat road the energy would all come from the fuel but on a steep slope it could all come from the change in height of the car. (Figure 1.)

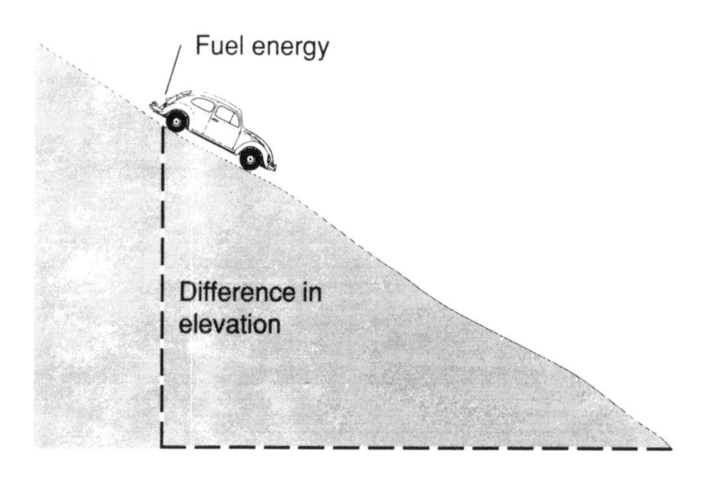

Figure 1

In the same way, to overcome the friction of the moving water against the pipe surface, water flowing along a pipe requires energy. Initially, it is all supplied by the change in altitude *(potential energy)* but it can be conver-ted into an internal energy store known as *pressure energy*.

If we connect a glass tube with an open top to the top of a pipe full of water under pressure, provided the pipe is tall enough, the water will rise up the tube until the pressure at the bottom of the tube produced by the weight of the water column is the same as the pressure in the pipe. This column of water is called the *pressure head.* (Figure 2).

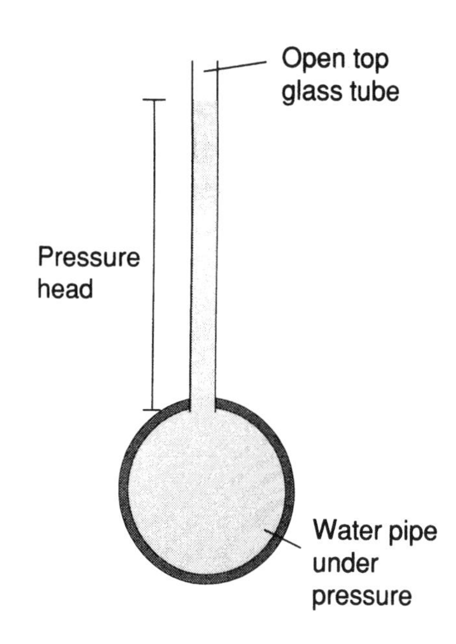

Figure 2

One big difference between energy in cars and pipes is that the energy in pipes can change the way it is stored. If we assume for a moment that the water can travel along the pipe without using any energy then some or all of the energy can be converted between pressure and potential energy. The total energy, however, remains the same. (Figure 3)

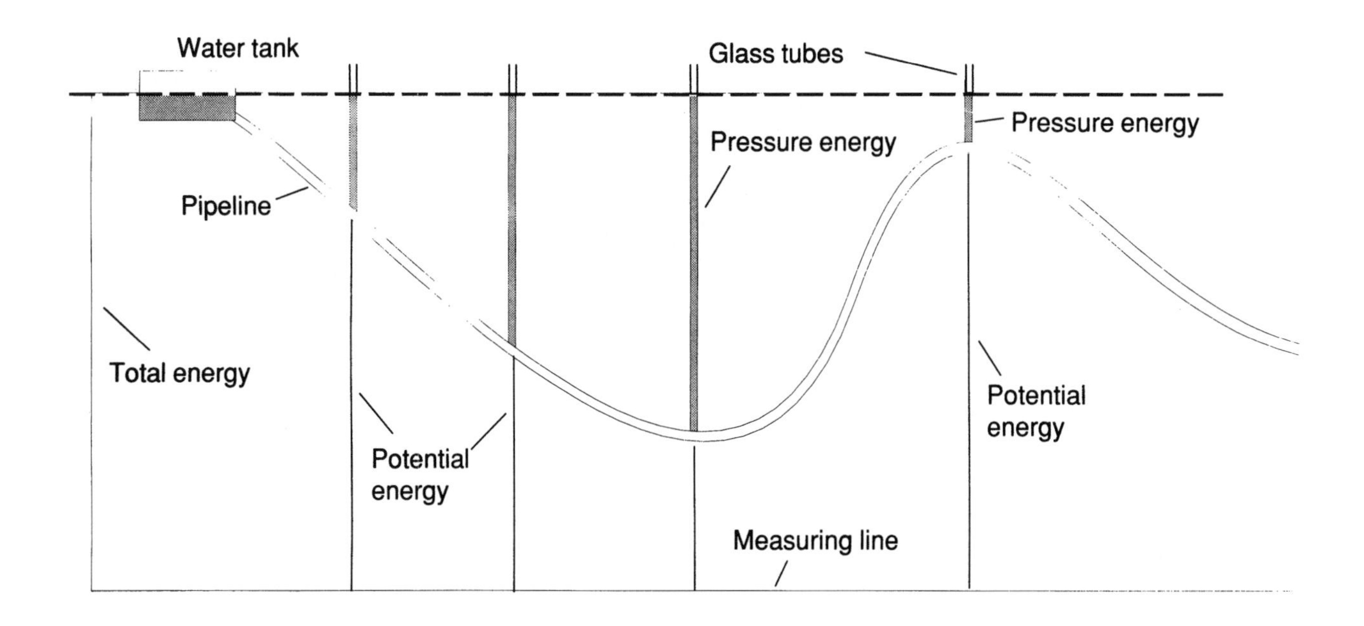

Figure 3

In practise energy is used up by friction as the water flows along the pipe such that the total energy available at the end is less than that at the beginning. This would be visible if we could fit a number of glass tubes in our pipeline because the water surface in the tubes would be seen to fall. If we could draw an imaginary line joining up the water levels in the glass tubes we would be able to measure the pressure in the pipe anywhere along the pipe and the difference in height between the line and another imaginary line drawn through the water surface level at the start of the pipeline would show the amount of energy lost. The sloping line is called the *hydraulic gradient line.* (Figure 4).

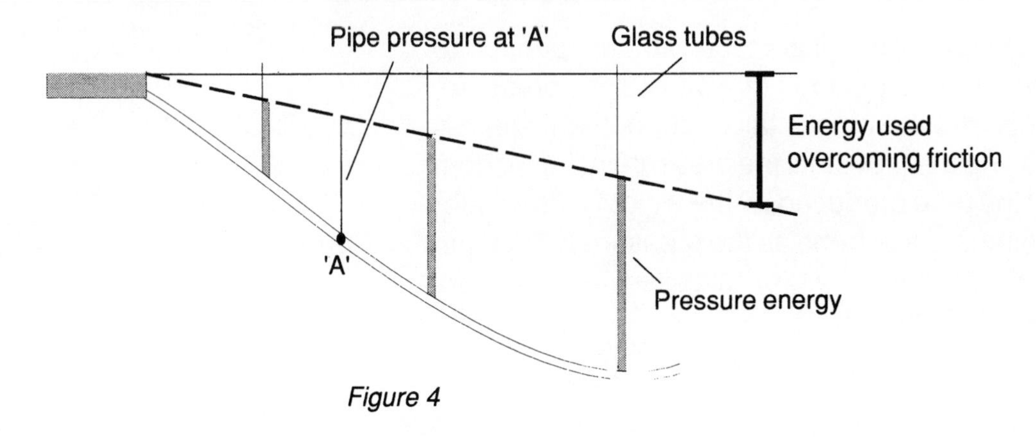

Figure 4

Designing the pipe

If we know how much energy we have at the start and end of a pipeline and we know the pipe length we can work out the slope of the *hydraulic gradient line.* Alternatively the total energy loss can be divided by the length of the pipeline to find the energy lost per metre length of pipe.

Table 1: Energy loss in metres per metre length of pipe

	PIPE MATERIAL AND DIAMETER IN MILLIMETRES												
Flow l/h	POLYTHENE (HDPE)				uPVC				GALVANIZED IRON (GI)				Flow l/h
	12	20	25	38	12	20	25	38	12	20	25	38	
500	0.18	0.01	0.00	0.00	0.21	0.01	0.00	0.00	0.28	0.02	0.00	0.00	500
1000	0.63	0.05	0.01	0.00	0.75	0.06	0.02	0.00	1.09	0.07	0.02	0.00	1000
1500	1.32	0.11	0.03	0.00	1.62	0.12	0.04	0.00	2.43	0.16	0.05	0.00	1500
2000	2.21	0.18	0.06	0.00	2.82	0.21	0.07	0.00	4.29	0.29	0.09	0.01	2000
2500	3.32	0.28	0.09	0.01	4.33	0.32	0.10	0.01	6.68	0.45	0.14	0.01	2500
3000	4.63	0.39	0.13	0.01	6.16	0.45	0.15	0.01	9.59	0.64	0.20	0.02	3000
3500	6.14	0.51	0.17	0.02	8.31	0.61	0.20	0.02	13.02	0.87	0.27	0.03	3500
4000	7.85	0.65	0.22	0.03	10.77	0.78	0.25	0.03	16.98	1.13	0.35	0.04	4000
4500	9.74	0.80	0.27	0.03	13.56	0.98	0.32	0.04	21.47	1.43	0.44	0.05	4500
5000	11.84	0.97	0.33	0.04	16.67	1.20	0.39	0.04	26.47	1.76	0.54	0.06	5000
5500	14.13	1.16	0.39	0.05	20.09	1.44	0.46	0.05	32.01	2.12	0.65	0.07	5500
6000	16.64	1.36	0.46	0.06	23.84	1.71	0.55	0.06	38.06	2.52	0.78	0.08	6000
6500	19.31	1.57	0.53	0.07	27.90	1.99	0.64	0.07	44.65	2.96	0.91	0.10	6500
7000	22.18	1.80	0.61	0.08	32.28	2.30	0.73	0.09	51.75	3.43	1.05	0.11	7000
7500	25.23	2.04	0.69	0.09	36.98	2.63	0.84	0.10	59.38	3.93	1.21	0.13	7500
8000	28.47	2.30	0.77	0.10	42.00	2.98	0.95	0.11	67.54	4.47	1.37	0.15	8000
8500	31.91	2.57	0.86	0.11	47.34	3.35	1.07	0.13	76.21	5.04	1.55	0.17	8500
9000	35.33	2.85	0.96	0.12	52.99	3.74	1.19	0.14	85.42	5.64	1.73	0.19	9000
9500	39.33	3.15	1.06	0.14	58.97	4.16	1.32	0.16	95.15	6.28	1.93	0.21	9500

***Note:** Pipes are assumed to be a few years old working under normal conditions*

Knowing the energy loss per metre length and the amount of water required it is possible to calculate the most appropriate size of pipe to use. This calculation is difficult to do by hand and so Table 1. gives a selection of useful results. The table shows the energy loss per metre length of pipe for different pipe sizes and different flow rates. The table is also divided into different pipe materials since these too have an effect on the amount of energy used. The best way to explain how to use the table is to give an example.

Example

It is hoped to use a spring 2 000m from a village as a source of water. What size of pipe must be laid if a flow of 3000 litres per hour is required and the spring is 50 metres higher than the village? (Figure 5 shows a section along the proposed pipeline.)

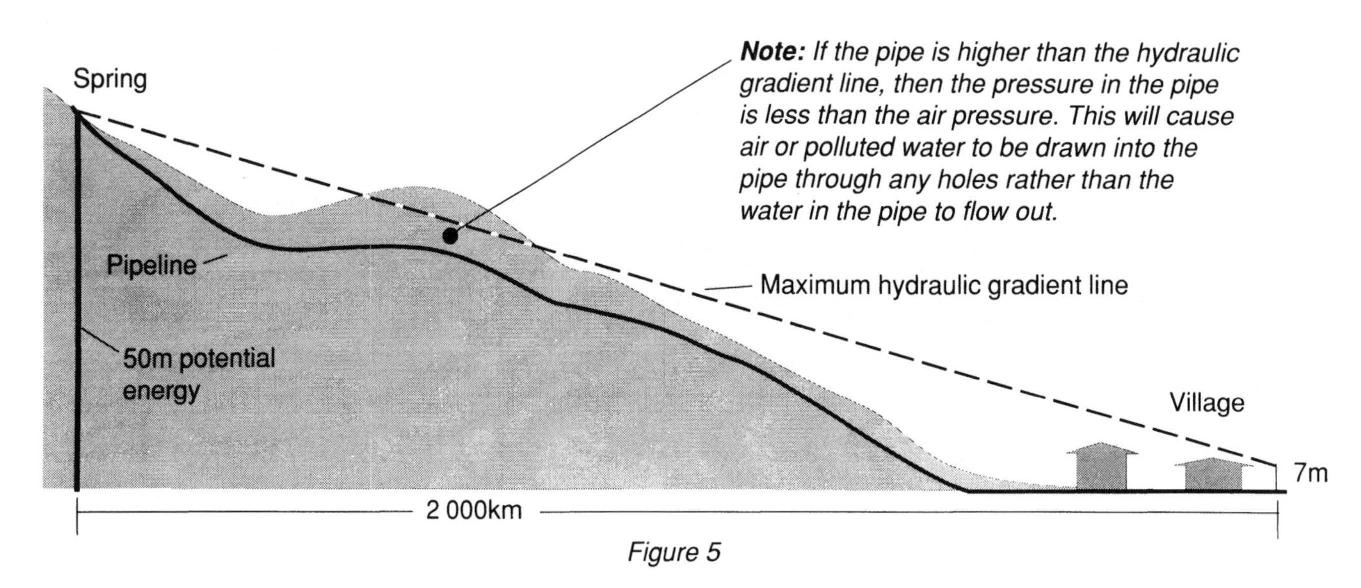

Figure 5

Answer

Energy available at the spring box	50m (potential energy)
Energy needed in the pipe in the village for distributing the water around the pipes in the village and out of the standposts	7m (pressure energy)
Maximum energy that can be lost by friction	50 - 7 = 43m
Energy loss per metre length of pipe (It is acceptable to use the pipe length if the horizontal length cannot be measured)	$\frac{43}{2000}$ = 0.02 m/m

From the table, for flows of 3 000 litres per hour we would need 38mm diameter for all pipe materials. This is the maximum flow for a 38mm GI pipe having an energy loss of 0.02m/m but HDPE and uPVC pipes of this diameter would provide 3500 l/h. The selection would depend on the cost, types of pipe available and type of ground in which the pipe was to be laid.

Note: *The pipe chosen must always have an actual energy loss less than the maximum allowable loss. Also, the maximum working pressure should be lower than that recommended for the piping used. Technical Brief No. 26: Public standposts gives head losses for pipe fittings and taps.*

Further reading:

Jordan Jnr, Thomas D., *A Handbook of gravity-flow water systems*. Intermediate Technology Publications, 1984.

Text: Bob Reed Graphics: Rod Shaw
WEDC, Loughborough University of Technology, Loughborough, Leicestershire LE11 3TU, UK.

30. Community management

In many countries significant numbers of water supply points and sanitation systems are out of action. The reasons for these failures were originally assumed to be due to use of inappropriate technology. Now planners and technologists are aware that many problems also arise because the consumers of these services, that is the local community, have not been sufficiently involved in the design, implementation, operation and maintenance of their own water supply and sanitation facilities. It is recommended that for rural and low-income communities the *Top Down* method must be replaced by the *Bottom Up* approach.

Four reasons for promoting community management:

- To maximize health benefits
- To ensure sustainability through effective operation and maintenance
- To ensure use of local resources, knowledge and skills so as to minimize costs
- To build up community confidence so as to enable further community development in other sectors

The different roles of participants in communit

COMMUNITY		
Felt need for improved water supply and sanitation Exposure to health education	Response to questions by health workers and agency/government staff about health, wealth, water and sanitation. Discussions regarding experiments into affordable means of improving water and sanitation.	Training of community members to assist with the programme. Training of local artisans and contractors

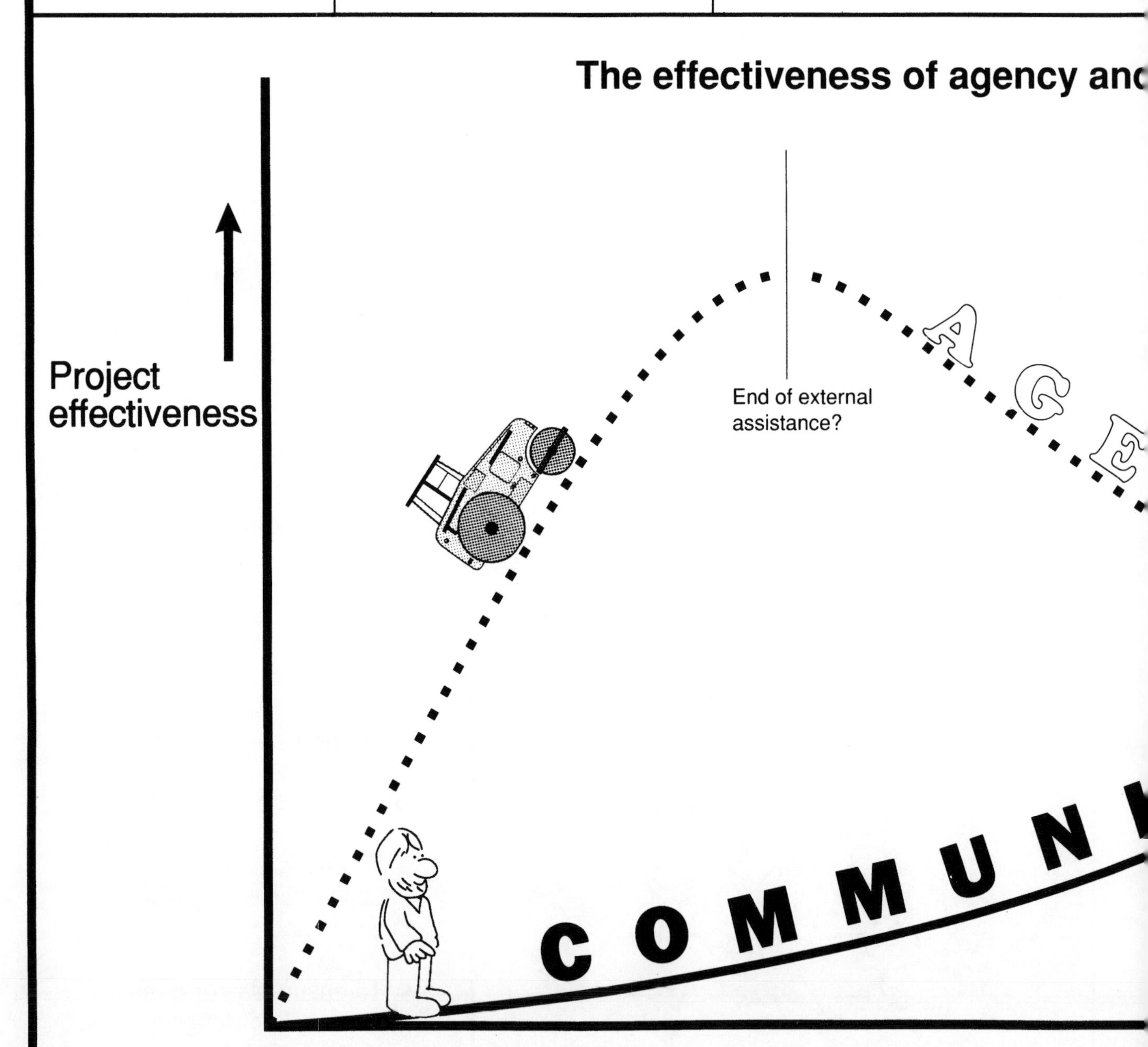

IMPLEMENTING AGENCY/GOVERNMENT	
Demonstration: Assignment of responsibilities; encouragement of health education; stimulation of demand; technical and social surveys; planning with communities and households; local testing of practical aspects of systems; establishing confidence of government and potential donors; training of field staff.	**Consolidation:** Integration with other government programmes (health, education, agriculture etc.); standardization of technical details; bulk ordering of materials with long delivery dates; establishment of revolving fund facilities; training of community development staff

FUNDING AGENCY/ GOVERNMENT		
Sector plan identification	Pre-feasibility	Feasibility, appraisal and approval, detailed design

water supply and sanitation programmes

	Publicity about the programme Visits by community representatives to working systems Drawings and models and cost forecasts made available to all Systems for financial assistance established COMMUNITY/HOUSEHOLD DECISION TO PARTICIPATE	Use and care of facilities	Comments about further improvements and upgrading

ommunity managed projects?

TY MANAGED

NCY MANAGED

?

Time

	Expansion: Mass promotion in the target communities; continued health education; use of mass media for information and selling; demonstration units as water and sanitation supermarket with financial, material and technical assistance where required; waiting for communities to respond; advice on responsibility of community and households for care and maintenance.

	Implementation	Operation and maintenance	Evaluation

Community management

There are a whole range of approaches by which implementing or facilitating agencies become involved in the provision of community water supply and sanitation. At one extreme the government or agency uses the potential consumers, that is the community, simply as unpaid labour. The agency plans and designs the system in a distant office, and then directs the people when and where and how to dig and build the system. This is sometimes described as 'Directive' community participation. At the other extreme the community decides that it wants to improve its water supply and/or sanitation. It takes responsibility for those improvements, requesting assistance from a facilitating agency as required. It uses the technical help that is offered and ensures that community members receive training in all necessary construction, operation and maintenance skills. The community mobilizes its own finance, any available external finance and plans how to pay for long term operation and maintenance. To the agency this could be called 'supportive' participation because they are supporting the community, not directing them.

Most projects lie between these two extremes but the most effective always try to follow the 'supportive' approach - for it is only through the supportive approach that 'ownership' of the project lies with the long term users. It is only through this goal of 'community management' that the long term benefits of improving water supply and sanitation are realized.

Five conditions for community success:

- Communities are involved in all stages of their water and sanitation projects.
- Roles and responsibilities of community and government and agencies are clearly defined and obligations are fulfilled.
- Government and agencies act as a supporter of the community, not as owner or manager of the water and/or sanitation system.
- Contact between community and agency is through staff whose primary skills are organizing and motivating communities.
- Government and agencies fulfill their limited but vital tasks of motivation, training and technical assistance.

Go to the people
Live among them

Learn from them
Love them
Start with what they know
Build on what they have:
But of the best leaders
when their task is accomplished
their work is done
the people all remark
'We have done it ourselves'

Tao To Loa Tzuching (700 BC)

For further information:

Briscoe J. and de Ferranti D., *Water for rural communities*, World Bank, 1988.
White A., *Community participation in water and sanitation*, IRC 17, 1981.
Oakley P. and Marsden D., *Approaches to participation in rural development*, ILO, 1984 .

Text: Richard Franceys Graphics: Rod Shaw
WEDC, Loughborough University of Technology, Loughborough, Leicestershire LE11 3TU, UK.

31. Latrine vent pipes

Ventilated improved pit (VIP) latrines are recommended for unsewered communities in Africa and elsewhere - wherever solid waste material is used for anal cleansing.

VIP latrines take various forms:

Ventilation is provided in VIP's by a vent pipe with flyproof netting at the top.

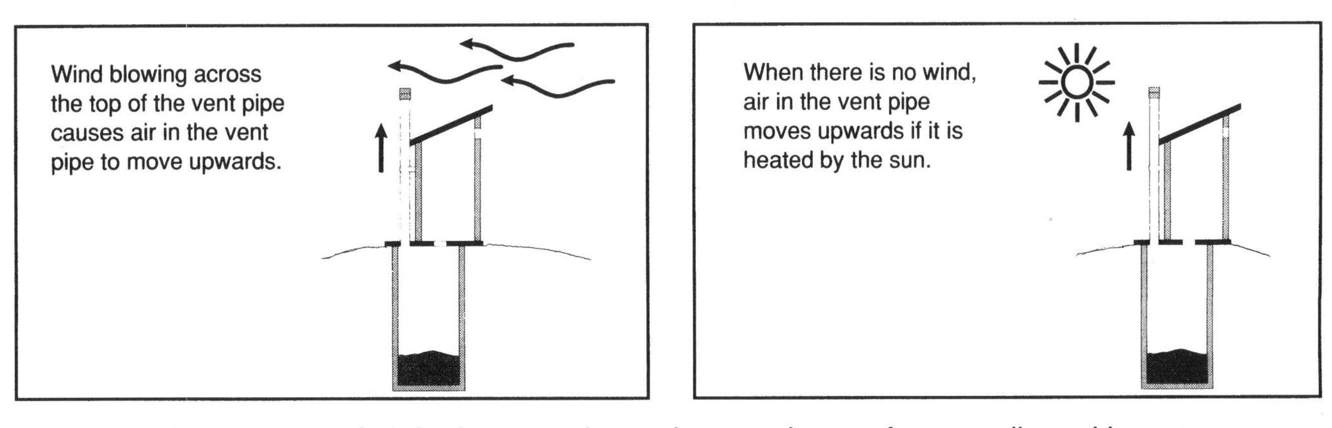

The upward movement of air in the vent pipe reduces nuisance from smells and insects.

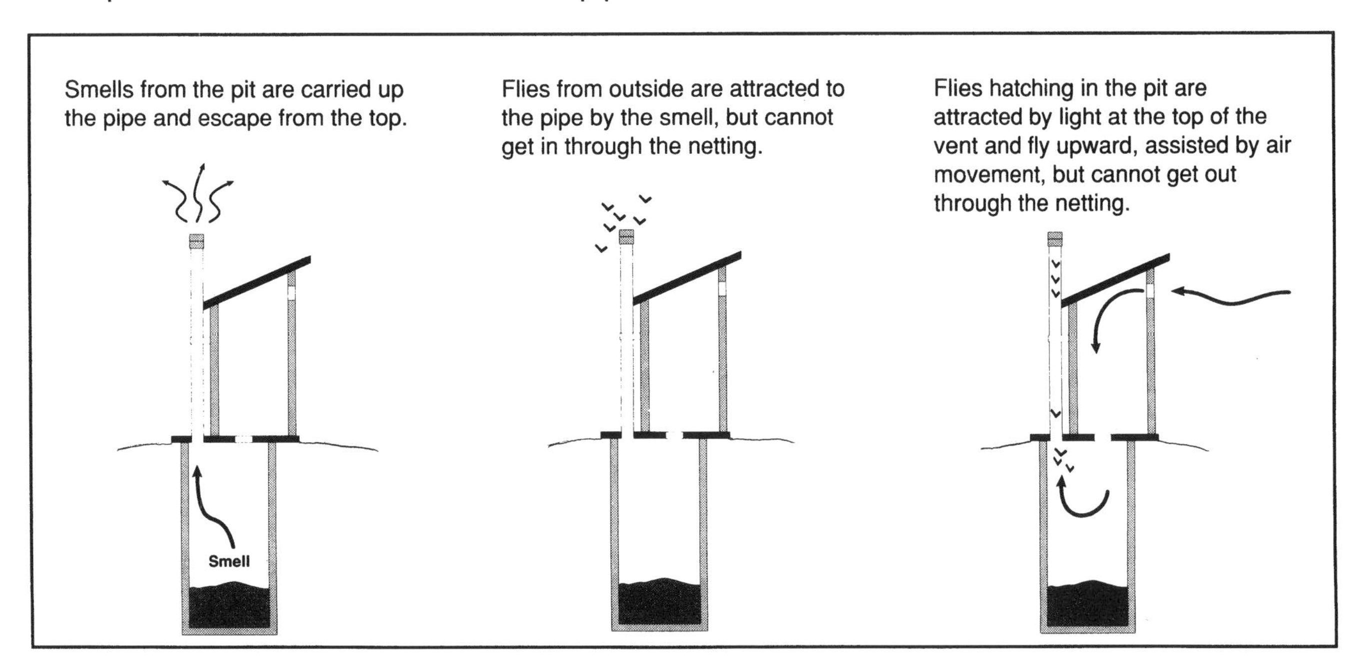

Materials for vent pipes

Asbestos cement pipes
PVC and uPVC pipes

PVC pipes should preferably have a stabilizer to prevent damage by ultra-violet light

Minimum internal diameter

In areas with high wind speeds: 100mm
Latrines built at minimum cost: 100mm

In areas with low wind speeds: 150mm

Multiple pit latrines where each pit is used by two cubicles: 200mm

Brickwork or blockwork

A vent can be built as an extension of the superstructure.

It can be inside or outside the building.

Inside vents must not make the latrine uncomfortable to use.

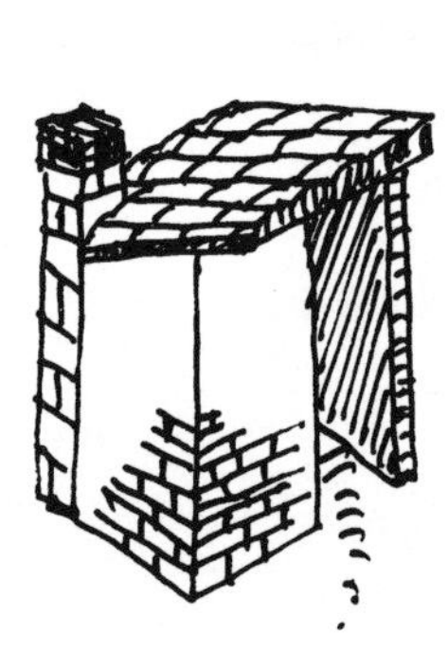

Minimum size

In areas with high wind speeds: 180mm square inside

Latrines built at minimum cost: 180mm square inside

In areas with low wind speeds: 230mm square inside

Locally made vent pipes

Large diameter bamboo
Remove all dividers

Plastered sackcloth on steel mesh
Pipes are made as follows:

- Cut a piece of strong steel mesh 2.5m long and about 0.8m wide (suitable mesh is a spot-welded 4mm bars at 100mm centres).
- Roll the steel mesh into a tube.
- Stitch sackcloth (hessian) tightly round the steel mesh tube
  Optional: - *Make a horizontal bath, for example, by cutting a 200-litre oil-drum lengthwise and welding halves together.*

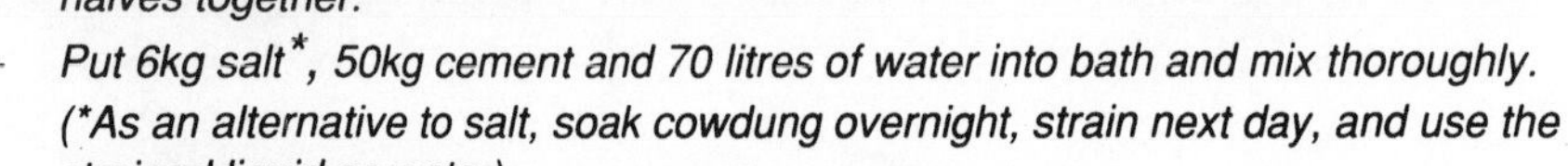

  - *Put 6kg salt*, 50kg cement and 70 litres of water into bath and mix thoroughly. (*As an alternative to salt, soak cowdung overnight, strain next day, and use the strained liquid as water).*
  - *Roll tubes slowly in the bath until all sacking is well soaked.*
  - *Keep pipe moist for four days, then allow to dry.*
- Plaster outside of tube with thin layers of mixture of sand, cement and water (for example, 2 parts of sand, 1 part of cement and enough water to make the mixture like thick soup that can be applied with a brush).
- Brush on more layers of plaster until total thickness is at least 10mm, taking care not to put plaster on the flyproof netting

Plastered matting - see opposite page

Anthill soil - see opposite page

Minimum internal diameter

In areas with high wind speeds: 200mm

In areas with low wind speeds: 250mm

More about locally-made vent pipes

Plastered matting

Straight reeds, bamboo or wood poles about 10mm diameter are tied together with wire or string to make a mat 2.5m long by 1m wide.

- Roll the mat around green saplings to make a tube about 300mm diameter.
- Fix flyproof netting to one end.
- Lay on ground and plaster half the tube with a layer of cement mortar (one part cement, three parts sand, not too much water). Keep moist for four days, then allow to dry.
- Fix pipe to latrine wall, plastered part against wall.
- Plaster the other half.

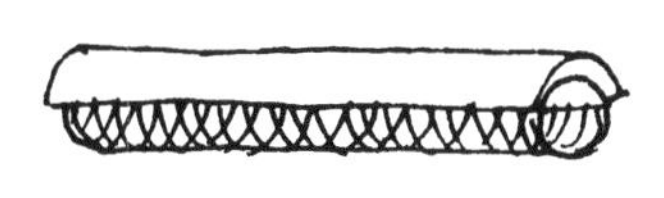

Anthill soil

- Knead anthill soil (like kneading dough to make bread).
- Make into large sausages about 100mm diameter and 900mm long.
- Make a sausage into a ring, which will be about 200mm inside diameter.
- Place ring in hole left in pit slab for the vent pipe
- Drive short lengths of reed or thin bamboo vertically into the ring.
- Add another ring on top. Drive in reed or bamboo and continue to height of 2.5m.
- Fix flyproof netting at top.
- Smooth outside of pipe
- Apply a thin layer of cement mortar (1 part cement, 6 parts sand) to outside.

Flyproof netting

Size of mesh

The best size of mesh is 1.5mm x 1.2mm

Larger holes allow flies to get through.

Smaller holes restrict air flow.

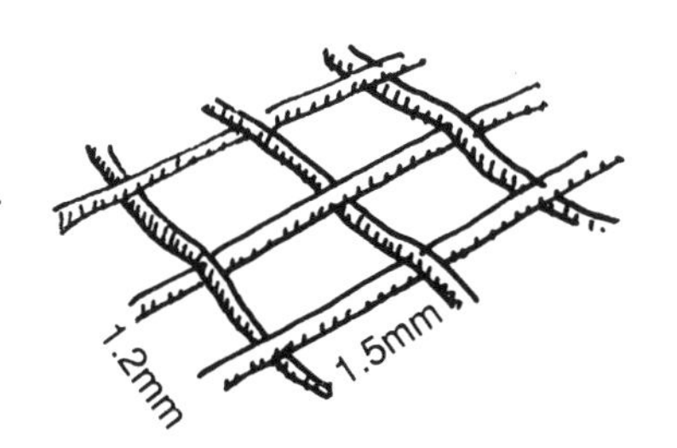

The best material for flyproof netting is **PVC-coated glass fibre**, which lasts more than five years.

Cheaper material may fail because of corrosion and attacks by birds and small animals.

Stainless steel lasts longer but is very expensive.

Fixing flyproof netting

For PVC or AC pipes sandpaper top of pipe so there is no sharp edge that will cut the netting

Fix to pipe with spray resin glue or tie round with galvanized wire or nylon string.

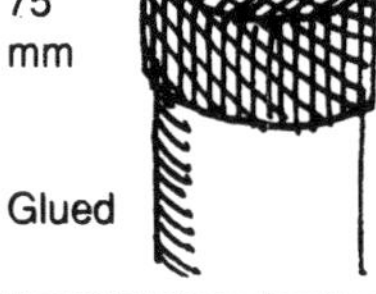

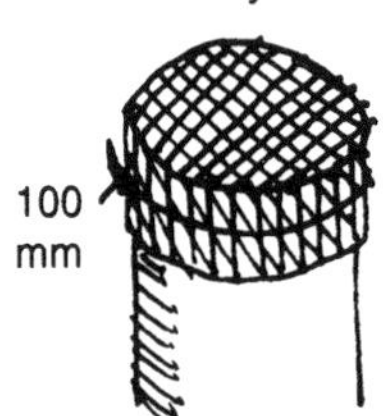

For bricks or blocks either build in or fix with pieces of wood.

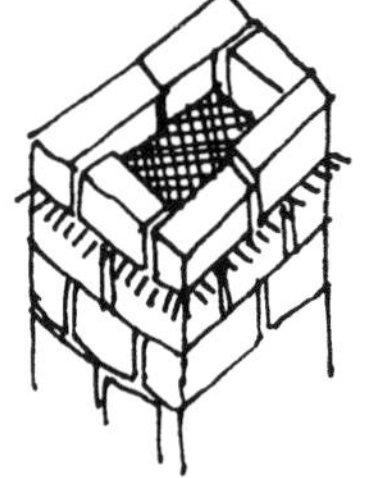
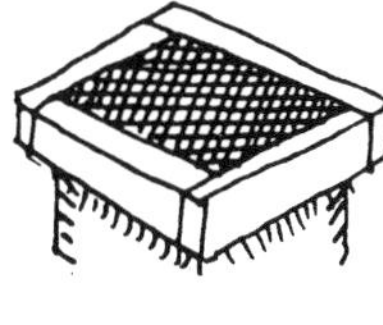

For plastered sackcloth, sew netting to sackcloth before plastering

For plastered matting fix netting to matting tube with galvanized wire or nylon string before plastering

For anthill soil fix netting below the top 'sausage'

Flyproof netting

Where upflow of air depends on the wind, the latrine doorway should face the direction of the prevailing wind.

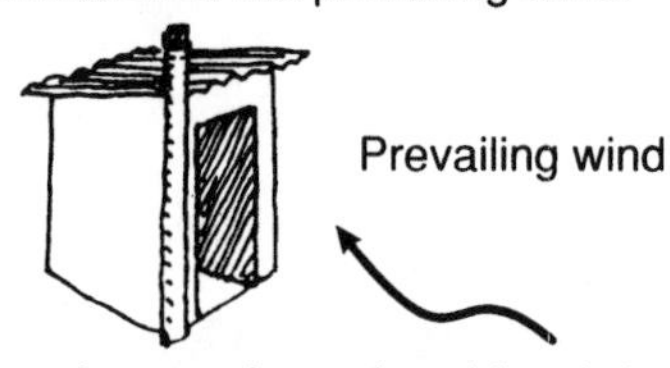

If the pit extends to the side of the superstructure the vent pipe should also face the prevailing wind.

A spiral latrine can easily be located so both opening and pipe face the prevailing wind.

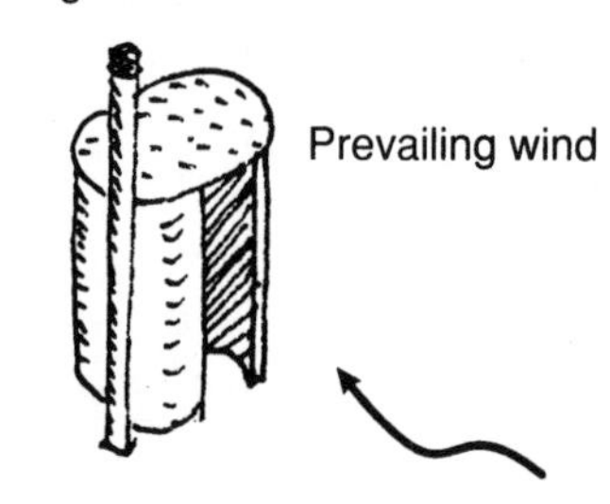

If the pit extends behind the superstrucure the doorway should have the wind; the vent pipe is opposite the wind.

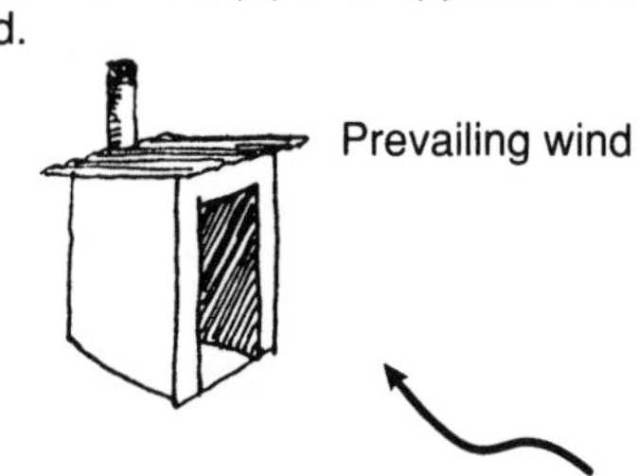

Fitting vent pipes

The bottom of the vent pipe should be securely fixed over a hole in the pit cover slab.

Cement mortar fillet

A PVC or asbestos cement pipe can be lowered into a socket set into a concrete slab.

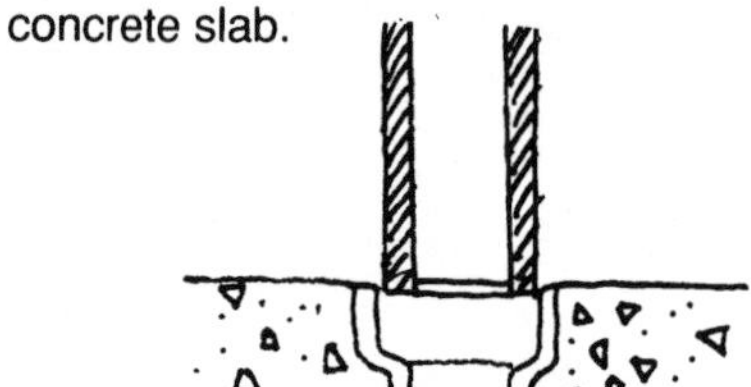

The pipe should be attatched to the wall of the superstructure with steel straps or galvanized steel wire built into the wall.

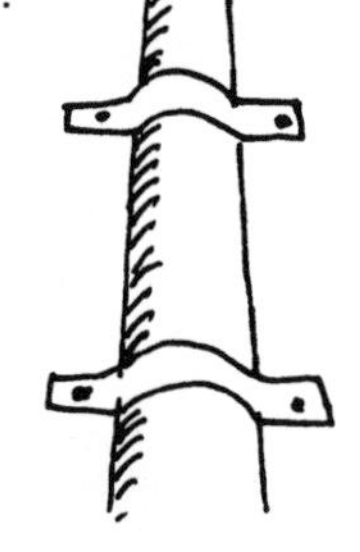

Inspection and maintenance

Inspect flyscreen regularly (at intervals of six months or less)

- Clear any debris from the screen, for example by pouring a bucketful of water down the pipe; this will also wash spiders and spiderwebs into the pit.
- Check the fixture of the vent pipe to the structure and replace if damaged.
- Make sure the vent pipe is sound and is firmly fixed to the slab.

For further information:

Franceys, Pickford, Reed. *On-site sanitation,* WHO, 1989.

Text: John Pickford Graphics: Rod Shaw
WEDC, Loughborough University of Technology, Loughborough, Leicestershire LE11 3TU, UK.

32. Drainage for improved health

The objectives of drainage

The principal function of drainage is to remove unwanted water from an area as rapidly as possible. Good drainage is critical to the general well-being of a site. Lack of adequate drainage causes rapid deterioration of road and path surfaces, restricts pedestrian and vehicular movement, results in damage to buildings and their contents, and creates generally insanitary conditions including potential sites for insect breeding.

The requirements are for:

- **Drainage of sullage,** that is, household wastewater which has been used for washing, cooking or cleaning purposes, but which does not contain excreta;

- **Drainage of stormwater,** that is, water which runs off the buildings and land as a result of rainfall.

Separate sullage drainage is not required if sewerage is used as the system of sanitation; all sullage can be discharged into the sewers.

Sullage drainage

It is important to ensure adequate sullage drainage both from houses and communal water supply points such as standposts and handpumps; between 50-80 per cent of the water supplied may end up as sullage. Water from personal use and clothes washing may be contaminated with pathogens, but to nothing like the same extent as toilet wastes. There is likely to be a significant amount of organic matter in water which has been used for food preparation and cleaning cooking utensils.

The quantity of sullage produced varies with the quantity of water supplied and local bathing practices. The provision of individual household water connections significantly increases the volume of sullage to be disposed of. The use of large quantities of water for bathing at communal standposts or wells can create highly insanitary conditions if the drainage is inadequate.

The problems resulting from inadequate disposal of sullage tend to be indirect, rather than due to the actual quality of the wastewater itself. Pools of sullage become breeding grounds for flies; the decay of organic matter may result in unpleasant smells; a generally insanitary environment results, in which certain pathogens, such as worm eggs, can survive.

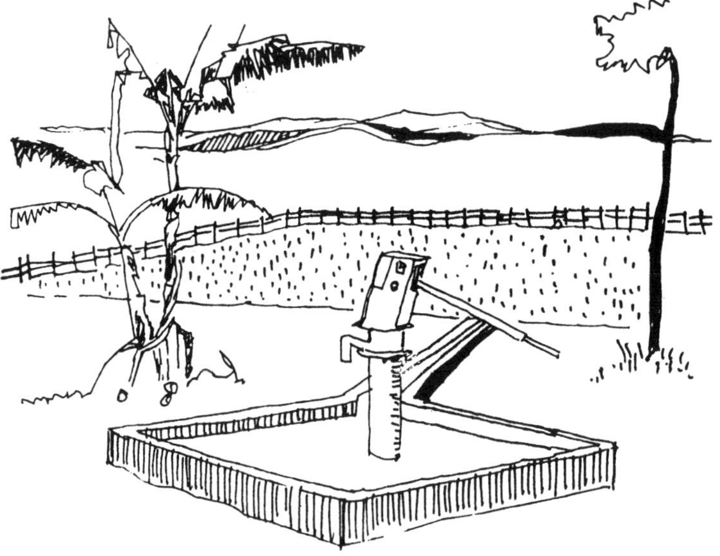

Figure 1. Water drained from water points can be used for production of fruit and vegetables. This can lead to improved nutrition and income generation.

On-plot disposal

Sullage can be disposed of within the housing plot, either by using the sullage for garden watering, (Figure 1) or by allowing it to percolate through the soil by means of a soakage pit as shown in Figure 2. The suitability of this method of disposal depends upon the quantity of sullage, the plot size, and the permeability the ground. If the ground is very sandy and highly permeable, it may be feasible to dispose of sullage into a latrine pit. Garden watering is only appropriate if plots are large; certain plants and trees, for example the banana tree, take up large quantities of water. On-plot disposal may be feasible where water is being fetched from a public water supply point. However, it is unlikely to be appropriate when the houses have individual water connections unless the ground is very sandy.

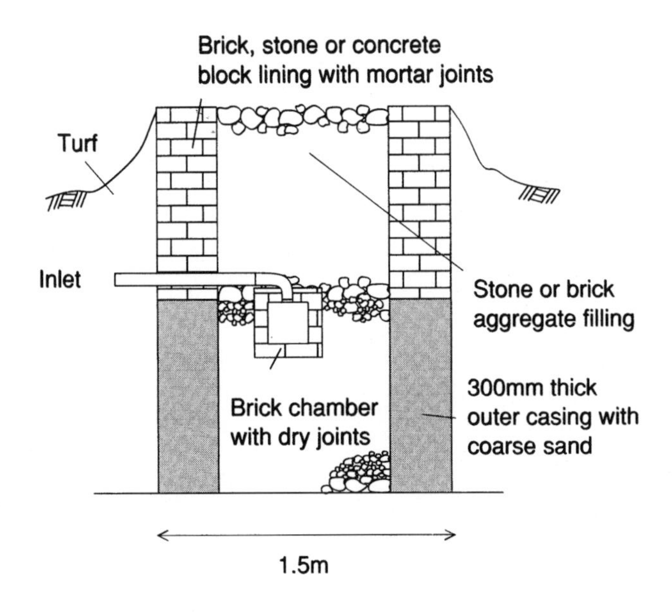

Figure 2. Soakage pit (after Indian Standards Institution)

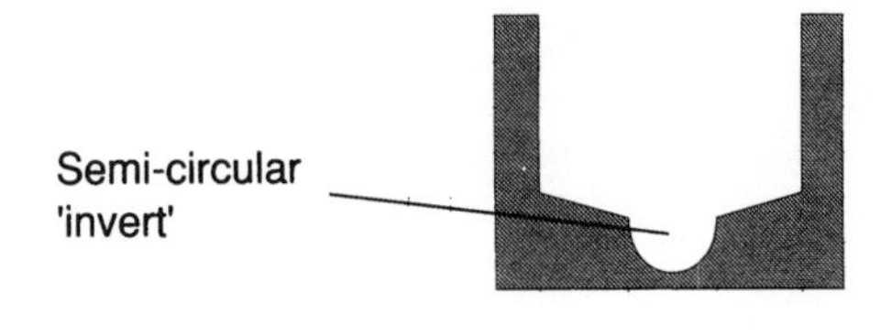

Figure 3. Compound section of a lined open drain

Use of Stormwater drains

Sullage can be discharged into the stormwater drains; problems may arise due to suspended matter settling out in the drain invert and careful hydraulic design is required to avoid this. Lined open channels having a compound section should be used wherever possible.

Stormwater drainage

Rain which falls on firm impervious surfaces such as roads and the roofs of buildings will run off that surface without being absorbed and the stormwater drainage system must have the ability to remove that water. There should normally be a drain running alongside roads and pathways which collects the rainwater from the road surface and surrounding buildings (Figure 4).

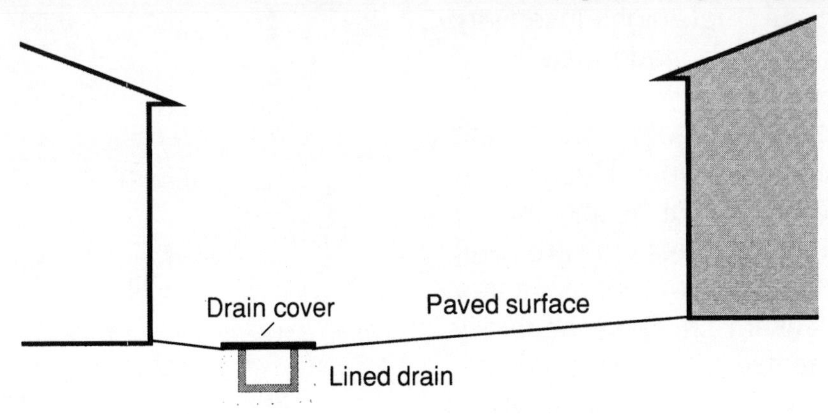

Figure 4. A roadside drain

Water in the drains should flow by gravity; it is therefore important that drains slope downhill in the direction of flow. The drain will normally slope at a similar gradient to the ground; if the ground is very flat, the recommended minimum gradients are 1:300 if the drain carries only stormwater and 1:150 if the drain carries sullage.

The drainage of large built-up areas in towns and cities is complex and needs to be designed by a qualified engineer.

Open channel drainage networks

These are relatively simple to construct and maintain, but take up space and pose a hazard to road users, especially if the drain is very wide or deep, or passes along a busy thoroughfare. In such cases the drains can be covered with removable slabs. The simplest open channel drain is a hand-dug, unlined ditch (Figure 5). Although there are limitations on its use, they are usually much cheaper than open channels lined with masonry or concrete. However lined drains (Figure 6) require less maintenance as they do not suffer from erosion; regular cleaning to remove blockages and debris is vitally important.

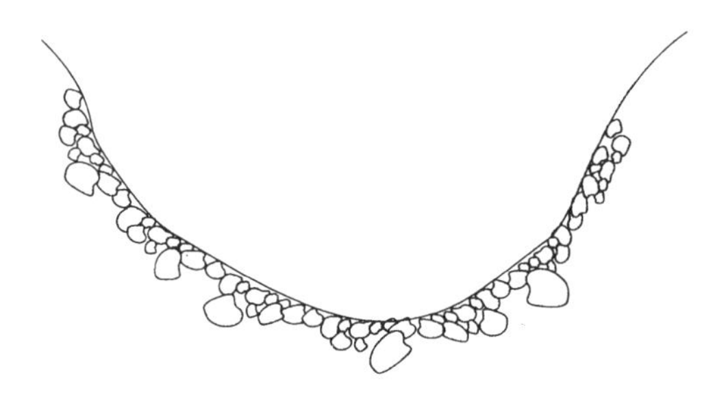

Figure 5. An unlined drain

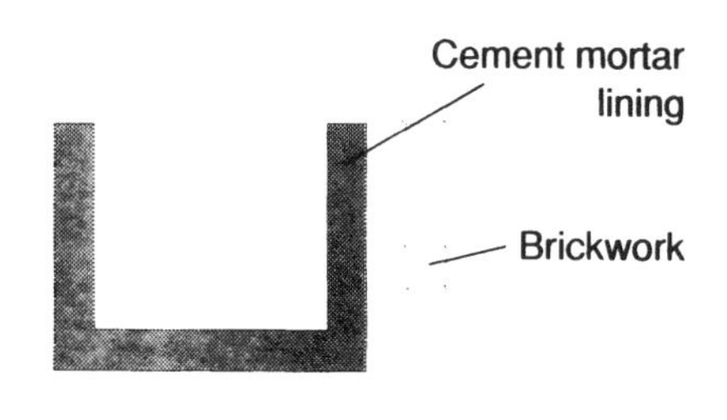

Figure 6. A lined drain

If streams or ditches which carry the drainage water from other areas pass through the site, improvement of the channel section may be necessary to prevent the bed and banks from eroding during high flows (Figures 7 and 8).

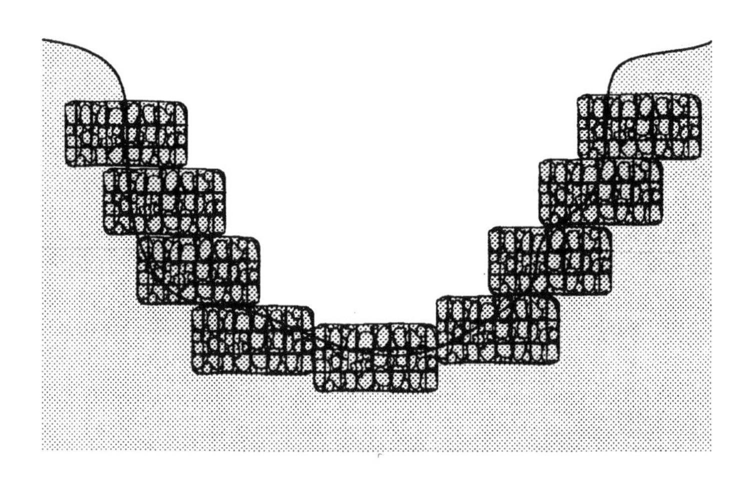

Figure 7. Improvement of a channel using gabions

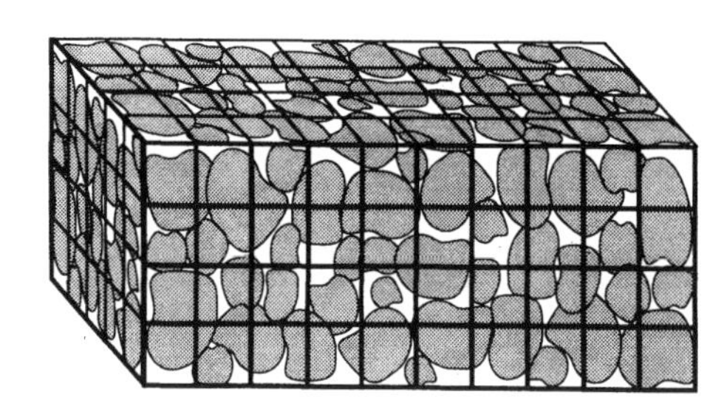

Figure 8. A gabion is a pack of stones or rocks held together by a wire mesh

Road-as-drain

In some densely populated settlements, paved roadways and alleys are used to carry stormwater short distances to drainage channels; that is, water is deliberately allowed to flow along the paved surface and there are no channels alongside. This works where the surfaces are fully paved and well maintained; it is only applicable if adequate sullage disposal facilities exist and in general is not recommended other than for small, fully paved areas (Figure 9).

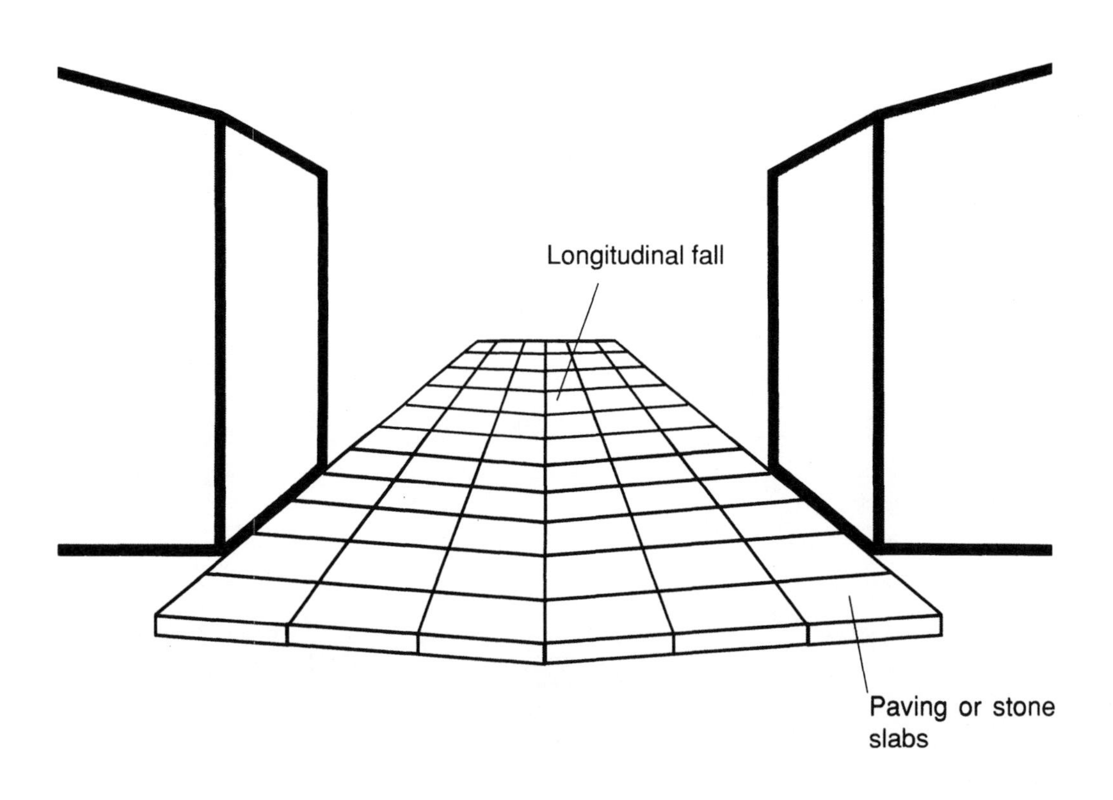

Figure 9. In some areas paved roads are used for drainage.

For further reading:
UNCHS, *Community participation and low-cost drainage.*Training Module , 1986, Nairobi, Kenya.
WEDC, *Services for urban low-income housing.* Fifth edition, 1988, Loughborough, UK.
Indian Standards Institution, *Code of practice for design and construction of septic tanks.* IS 2470, 1968 New Delhi, India

Text: Andrew Cotton Illustrations and design: Rod Shaw
WEDC, Loughborough University of Technology, Loughborough, Leicestershire LE11 3TU, UK.